T0275969

SpringerBriefs in Applied Sciences and Technology

More information about this series at http://www.springer.com/series/8884

Johannes Unger · Marcus Quasthoff
Stefan Jakubek

Energy Efficient Non-Road Hybrid Electric Vehicles

Advanced Modeling and Control

 Springer

Johannes Unger
Technische Universität Wien
Vienna
Austria

Stefan Jakubek
Technische Universität Wien
Vienna
Austria

Marcus Quasthoff
Intellectual Property and Innovation
  Management
Liebherr Machines Bulle SA
Bulle
Switzerland

ISSN 2191-530X   ISSN 2191-5318   (electronic)
SpringerBriefs in Applied Sciences and Technology
ISBN 978-3-319-29795-8   ISBN 978-3-319-29796-5   (eBook)
DOI 10.1007/978-3-319-29796-5

Library of Congress Control Number: 2016931601

Printed on acid-free paper

This Springer imprint is published by SpringerNature
The registered company is Springer International Publishing AG Switzerland

Preface

As already realized by the general community, the electrification of passenger cars has increased significantly over the past ten years, starting with micro-hybrid cars, which provide start-stop systems and limited energy recovery from braking, and going over to mild and full-hybrid electric vehicles, as well as full electric cars. This trend of growing electrification also occurs in the segment of heavy-duty vehicles and long-haul trucks.

The sociopolitical pressure for higher energy efficiency of on-road vehicles will also be imposed on non-road mobile machinery in the foreseeable future. Increasing energy prices will also enhance the economical motivation to go for higher energy efficiency. Any additional costs due to the energy efficiency increase of mobile machinery, which are usually counted as assets, need to be compensated by reduced fuel consumption or increased productivity, though. An environmental sustainability life cycle analysis requires therefore to include the resourcing, production, results of application, and the component disposal after lifetime to see if the energy efficiency increase is reasonable by ecological aspects.

Therefore, the conclusion to achieve a positive life cycle leads to the fact that any additional components of the hybrid powertrain and their degrees of freedom are used in the best way. This goal is the key topic of the book!

In comparison to on-road vehicles, the spectrum of load profiles of the powertrain is significantly larger and more dynamic for mobile machinery applications. The usual load profile is not only obtained by the drivetrain, but also by auxiliary equipment such as oil pumps, e.g., for lifting a bucket. On the other side, the total number of vehicles is much smaller and, therefore, development costs are more expensive.

A generic energy management, which is capable of controlling a wide range of totally different applications at an optimum, is presented to meet the requirements for non-road mobile machinery. Without any change to energy management software, the operation strategies such as torque split for maximum energy efficiency, phlegmatization, downspeeding, constant speed, and stationary operation may be changed between each other for flexibility reasons.

Cycle detection and load prediction, as presented in the book, provide further information for energy management, which help to operate the hybrid powertrain more on the optimal operation point and closer on the feasibility limit. In this context, the battery charging status needs to be known with high accuracy, but is not measurable online during operation. For this purpose, a generic methodology has been developed to identify battery dynamics automatically; it is independently applicable to any battery chemistry.

This book discusses several different sectors of technology and highly sophisticated engineering, but it should not be seen as a general reference or as a fundamentally educational book. It is more a book of detailed research content that can be used for implementation into given problem sets as well as a reference for different methodologies applicable to any other fields.

The book's content resulted from a project that had been financed by the Austrian Research Promotion Agency (FFG), and which was overseen by Graz University of Technology. The project consortium consisted of Graz University of Technology, Vienna University of Technology, Liebherr Machines Bulle SA, Liebherr Werk Bischofshofen, Liebherr Werk Nenzing and Kristl, Seibt & Co GmbH.

We, the authors, want to direct special thanks to Prof. Helmut Eichlseder, Prof. Martin Kozek, Dr. Peter Grabner, Dr. Herbert Pfab, Dr. Wilfried Rossegger, Dr. Christoph Hametner, Dr. Christian Mayr, Dr. Oliver König, Dipl.-Ing. Michaela Killian, Dipl.-Ing. Wolfgang Monschein, Dipl.-Ing. (FH) Christoph Kiegerl, Dipl.-Ing. Rupert Gappmaier and Mr. Hans Knapp for their help and inputs, which made this book possible.

Vienna/Bulle   Johannes Unger
December 2015   Marcus Quasthoff
                                                                                        Stefan Jakubek

Contents

Chapter 1
Introduction

Abstract Nonlinear system behavior of hybrid electric vehicle powertrains demands
advanced methodologies in modeling and control in order to achieve optimal results
by energy and battery management systems of such powertrains. This book describes
the main problems in the real-time control of parallel hybrid electric powertrains in
non-road applications, which work in continuous high dynamic operation. In order
to maximize the energetic efficiency of such powertrains, the operation point of the
engine must be kept in the optimal region. In terms of driveability, the optimal regions
are mostly in regions with lower dynamics, which leads to engine stalling at high
load peaks. This must be prevented by the controller in any case. For this purpose, the
book addresses an energy management control structure, which considers all constraints of the physical powertrain and uses novel methodologies for the prediction
of future load requirements to optimize controller output in terms of an entire work
cycle of a non-road vehicle. The load prediction includes a methodology for short
term loads as well as for an entire load cycle by means of cycle detection. This way,
energy efficiency can be maximized, which simultaneously results in a reduction of
fuel consumption and exhaust emissions. One significant information required by the
energy management system is the battery's state of charge, which is not measurable
on-line. This requires an accurate state of charge estimation, which is based on a
dynamic battery model. A novel methodology is introduced for the nonlinear battery
modeling that also considers the design of experiments of the measurements to identify model parameters with minimum variance. The reader of the book gets a deep
insight into the necessary topics to be considered in designing an energy and battery
management system for non-road vehicles and learns that only a combination of the
management systems can significantly increase the performance of a controller.

Keywords Non-road hybrid electric vehicle (HEV) · Model predictive control ·
Nonlinear system identification · Optimal model based design of experiments (DoE) ·
Load and cycle prediction

© The Author(s) 2016   1
J. Unger et al., *Energy Efficient Non-Road Hybrid Electric Vehicles*,
SpringerBriefs in Applied Sciences and Technology,
DOI 10.1007/978-3-319-29796-5_1

(a)   **(b)**

Fig. 1.1 Two classic examples of non-road mobile machinery. **a** Wheel Loader. **b** Excavator

1.1 Motivation

Non-road mobile machinery (NRMM), such as wheel loaders or excavators, is highly
dynamic, mostly cyclically used, applications (see Fig. 1.1) that are operated by
especially trained drivers to bring out the maximum performance of the vehicles,
c.f. Filla (2009). Usually, compared to on-road vehicles, higher power densities and
load dynamics occur, which increase the exhaust emissions of non-road vehicles significantly (Lin et al. 2003). In the past years, the legislative regulations for exhaust
emissions of such vehicles also became more stringent (e.g., *US EPA Tier 4* respectively *EU Stage IV*, (The European Parliament and Council of the European Union,
1997)) and will be more severe in the future (up to particle counting in, e.g., *EU Stage
V*, (The European Parliament and Council of the European Union, 2014)). Exhaust
after treatment systems (EATS) are in general used to keep the exhaust emission
regulations, but they are cost intensive and may not be enough for future regulations
(Weibel et al. 2014). Lowering the rotational speed of the powertrain (downspeeding)
and limiting the dynamics of the engine torque (phlegmatization) can decrease fuel
consumption and production of emissions in the engine, but these are contrary objectives compared to the required powertrain dynamics (Unger et al. 2015). Considering
the future load demand as well as cycle information in the control of the powertrain
might also have positive effects on exhaust emission reduction, whereby the load
demand of non-road machinery is mostly unknown and directly dependent on the
driver. Nevertheless, a reduction of the generated exhaust emissions not exclusively
achieved by EATS is desirable.

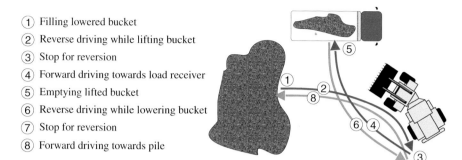

① Filling lowered bucket
② Reverse driving while lifting bucket
③ Stop for reversion
④ Forward driving towards load receiver
⑤ Emptying lifted bucket
⑥ Reverse driving while lowering bucket
⑦ Stop for reversion
⑧ Forward driving towards pile

Fig. 1.2 Classic Y-cycle of a wheel loader loading a truck with bulk material

1.2 Characteristic Applications of Non-Road Mobile Machines

Applications, where non-road mobile machines are established, are manifold. They are used in earth moving, material loading, refilling of bulk storages, ditching, building demolition, flattening, and many other applications, which demand different requirements on their powertrain. The requirements are basically defined by the dynamics of the powertrain loads and need to be provided by the powertrain without engine stalling at high dynamic load peaks. In terms of energy, the drivetrain of non-road vehicles is mostly the main energy consumer, which sometimes enables recuperation of regenerative power. The wheel loader is such an application that has high drivetrain energy consumption as well as high dynamic load requirements due to the rough environmental influences acting on the machinery during operation (e.g., unsuitable grounds, slopes or task impacts,…). Therefore, it is chosen as an example of the concepts and proposed methodologies that are presented in this book.

Bucket-equipped wheel loaders are very often used in the earth moving industry to transport bulk material or load trucks with it. The driver controls the vehicle with a joystick and an accelerator, which moves the shovel and requests the driving speed, respectively. Load trajectory and driving patterns are exclusively dependent on the driver and unknown in advance. Typical loading cycles—such as V or Y-cycles—are repeated periodically for a few times, and thereafter the operation is changed to another cycle. In Fig. 1.2, a Y-cycle is depicted exemplarily, which comprises several steps repeated periodically.

Filla (2013) analyzed different loading cycles to optimize the path trajectory in order to achieve reductions in fuel consumption and an increase in productivity. However, periodical operation is observable in the past load signal, which in fact can be considered for the control of the powertrain.

1.3   Configurations of Hybrid Electric Powertrains

Actual developments for conventional powertrains, which usually only consist of an internal combustion engine (ICE), may not be sufficient to reach the low emission and fuel consumption standards requested by law and customers, respectively. Therefore, a great deal of interest is given to hybridization of non-road vehicles, where the conventional fuel-based powertrain is enhanced with a secondary energy source to achieve lower fuel consumption and exhaust emissions, respectively. Commonly, an electric energy source such as an electrochemical battery or a double-layer capacitor (DLC) is used, though hydraulic concepts are available as well. Meanwhile, distinguished power capabilities are provided by power-type batteries, which make batteries—as compared to DLCs—more attractive to be used within non-road hybrid electric vehicles (HEV) (Unger et al. 2012a). Hybrid powertrains can be assembled in series or parallel configuration as well as in a combination of both, which is depicted schematically in Fig. 1.3.

Figure 1.3a depicts the series configuration, where no direct form-locking connection between the ICE and the load P_{load} is realized. A secondary energy storage module supplies and receives the power for the electric machines through an inverter unit (INV). Series configuration means to the electric motor (EM) that full power needs to be provided for the consumer, while the electric generator (EG) may have

Fig. 1.3 Typical hybrid powertrain configurations for hybrid electric vehicles. **a** Series, **b** Parallel, **c** Combination of series and parallel

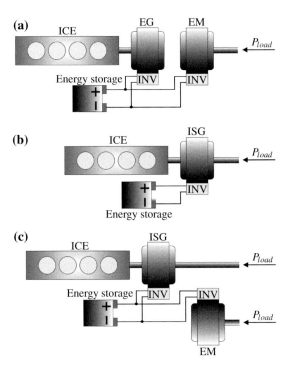

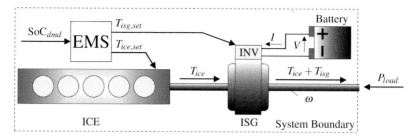

Fig. 1.4 Detailed schematic overview of a parallel hybrid electric powertrain used in non-road HEV

less power capabilities. In case of high energy conversions, the energy flow must be converted at least two times to reach the consumer, which is disadvantageous. In the parallel configuration (b), where ICE and EM are coupled form-locked on one axis, the rotational speed for the ICE and the integrated starter generator (ISG) is the same, while the torques add up. With this configuration, as long as no clutch is used between ICE and ISG, an electric only strategy cannot be realized without engine hauling. A combination of series and parallel configuration (c) has the advantage that electric power can be used for the auxiliary components. Much research interest has been spent on analyzing the different configurations (see e.g., Xiao et al. 2008, Bayindir et al. 2011, Katrasnik 2009), while following Kwon et al. (2010), the parallel configuration is most favorable for non-road vehicles, though. In Fig. 1.4, a parallel hybrid electric powertrain is depicted with a schematic overview of the energy management system (EMS). The depicted powertrain can be used in different non-road vehicles and applications, while the EMS needs to be especially parametrized for the corresponding vehicles.

The duty of the EMS is to keep the state of charge (SoC) of the battery at the demand value SoC_{dmd}. This can be achieved by controlling the ICE set-point torque $T_{ice,set}$ and ISG set-point torque $T_{isg,set}$ under consideration of all constraints of the components in such a way that the rotational speed ω is kept at an optimal value, while the unknown load P_{load} acts on the powertrain. Subsidiary component series controllers apply the set-points to the ICE torque T_{ice} and ISG torque T_{isg}. Battery current I and voltage V attune to T_{isg}, but must be kept below their limits to avoid physical damage. Due to the nonlinearities of the powertrain, achieving an optimal control performance amounts to a nonlinear optimization problem to be solved in the EMS in real time (Unger et al. 2015).

1.4 Challenges in Controlling Hybrid Electric Vehicles

Hybridization can enhance the degrees of freedom of the powertrain to provide the high dynamic load demand of non-road vehicles, but the engine needs to be limited in dynamics according to the downspeeding and phlegmatization strategies in order to reduce fuel consumption and exhaust emissions. The reduced dynamic tightens to keep all physical constraints of the system, though. Only the information of the entire future load cycle, which is in general unknown for non-road machinery, provides the possibility to reduce the dynamic as far as possible and to exploit the full energy storage capabilities of the battery (Mayr et al. 2011a). An electrochemical battery offers sufficient energy storage capabilities to recuperate regenerative power and consider motor braking phases resulting from an entire load cycle, but the disadvantage of batteries is that the SoC of the battery is not measurable online and requires an SoC estimator during operation (Plett 2004c). In this context, the SoC estimation is only accurate if a precise nonlinear dynamic battery model is used that is capable of the high dynamic loads occurring at non-road vehicles (Hametner and Jakubek 2013). Such model needs to consider the nonlinear dependency on SoC, temperature, and current, as well as many other nonlinear effects such as relaxation and hysteresis (Hametner et al. 2012).

Furthermore, as a result of hybridization, system nonlinearities are implicated by the electrical system, and the system complexity as well as costs is increased (Unger et al. 2014). Minimal costs can only be achieved if the powertrain design is matched with the application (Gao and Porandla 2005) and the decrease in fuel consumption and exhaust emissions is significant to reduce the operational costs of the powertrain as well as the acquisition costs for the EATS, respectively. The latter is only achievable if an appropriate framework is available that considers all information in the EMS. Such a framework is provided by a model predictive controller (MPC), which is an advanced method of process control that uses a model of the process to predict the future evolution of the process to optimize the control signal (Mayne et al. 2000). Due to the nonlinearities of the powertrain, a nonlinear optimization problem results within the MPC, for which real-time implementation is necessary in the hybrid control unit (HCU). However, fast sampling rates intensify the real-time requirements and predicting the entire future cycle load trajectory is difficult (Mayr et al. 2011a).

1.5   Proposed Concepts

This book presents methodologies for an efficient control of the described parallel hybrid electric powertrain that is based on the prediction of the future load trajectory as well as for the precise battery SoC estimation during high dynamic operation of a non-road vehicle, respectively. An electrochemical battery is chosen in this book in order to show the applicability of batteries as the secondary energy storage in non-road vehicles.

In the following, the functions of a battery management system (BMS) to precisely estimate the SoC are discussed first. Thus, a methodology for identification of high-accurate battery cell models is proposed, which is based on the architecture of local model networks (LMN). The global nonlinear model output is obtained by weighted aggregation of the outputs of dynamic local linear models, while the LMN structure is built by an automatic iterative algorithm. Nonlinearities of the battery cell are considered by corresponding inputs that provide sufficient information for the model. Since the LMN approach is only data-based, optimal model-based design of experiments is proposed to create optimal test sequences, which minimize the variance of the battery cell model's identified parameters. Based on an optimality criterion, which is obtained with the Fisher information matrix, a gradient-based algorithm is used for optimization, while constraints of the battery cell are considered to avoid physical damage of the battery cells. Note that due to the data-based approach, the methodology is applicable to different battery cell chemistries and also for DLCs. The obtained LMN battery cell model is then used to build the battery module model, for which a SoC estimator is built that is based on Kalman filter theory. Real measurements are made to verify the battery cell as well as module models and to show their cell chemistry independence. In order to fulfill the dynamic requirements of non-road machinery, a new high dynamic battery cell tester device was developed for the measurements of lithium-iron-phosphate and lithium-polymer battery cells. The real lithium-iron-phosphate battery module has been tested at a battery simulator/tester unit. A special measurement procedure, which is also proposed in this book, was followed exactly to achieve reproducible and comparable measurements.

Second, a real-time capable EMS is proposed, which is based on model predictive control; its primary objective is to minimize overall energy conversion, fuel consumption as well as exhaust emissions, while constraints are kept to avoid physical damage to the system (safety-related requirements) as well as to enforce that the degrees of freedom are optimally considered (efficiency-related requirements). The EMS consists of a cascaded controller architecture that refers to a linear slave and a nonlinear master MPC, respectively. In the prediction of the future evolution of the process in both MPCs, the unknown future load trajectory is considered by a prediction to increase the control performance as well as to achieve an optimal control during an entire load cycle. Thus, two methodologies are proposed for predicting short-term load peaks and detecting recurrent load cycles. Bayesian inference is used to statistically predict the short-term load based on the available powertrain signals such as driving speed or accelerator position, while recurrent load cycles are identified by a cycle detection (CD) that analyzes cyclic correlations within the past load trajectory by way of using the cross correlation function. Since the electrical system—including the ISG and the battery—is nonlinear, the optimization problem within the master MPC leads to a nonlinear optimization problem. A relaxation approach is used to solve the problem in real time, while simplifications are applied. Note that due to the small number of non-road applications, the EMS must be generically applicable to any non-road machinery to minimize development and implementation costs. In order to guarantee stability of the concept, stability as well as convergence are discussed.

Third, for the example of a wheel loader as representation of non-road machinery, the EMS is implemented on a real test bed to demonstrate the feasibility of the whole concept including accurate battery models. The obtained battery module model is implemented at the test bed battery simulator to emulate the battery during the measurements, which are made for different controller adjustments. Three main results are discussed in detail by means of simulation and real test bed measurements (Unger et al. 2015):

1. The feasibility of the proposed control concept with respect to the dynamic requirements of the machinery, when downspeeding and phlegmatization are applied.
2. The optimality of the control approach compared to the conventional powertrain by means of fuel consumption and emissions.
3. The benefit given by the cycle detection to exploit the full energy storage capabilities.

1.6   Main Contributions

This book is a revised and adapted version of the dissertation written by Johannes Unger (2015), where the authors' previously published papers form the main contributions.

Dissertation
JOHANNES UNGER: *Energy and Battery Management for Non-Road Hybrid Electric Vehicles*. Dissertation, Vienna University of Technology, 2015.

Paper A
M. QUASTHOFF, J. UNGER, S. JAKUBEK: *Entwicklungsmethodik eines generischen Batterie-Simulationsmodells und dessen Einsatzmöglichkeiten*. 5. Fachtagung Baumaschinentechnik 2012 in Dresden, Baumaschinentechnik 2012—Energie, Mechatronik, Simulation, Dresden, Schriftreihe der Forschungsvereinigung Bau- und Baustoffmaschinen e.V. (FVB), Heft Nr. 44, pages 263–284, 2012.

In Paper A, the focus lies on the identification of accurate battery models for the application in non-road machinery. The methodology of the local model network applied to battery modeling is given in detail and results are presented without consideration of the temperature in the model.

Paper B
J. UNGER, C. HAMETNER, S. JAKUBEK, M. QUASTHOFF: *Optimal Model Based Design of Experiments Applied to High Current Rate Battery Cells*. IEEE International Conference on Electrical Systems for Aircraft, Railways, Ship Propulsion and Road Vehicles (ESARS 2012 Edition), Bologna, ISBN: 978-1-4673-1371-1, pages 1–6, 2012.

In Paper B, the optimal design of experiments for battery cells is presented in detail, without consideration of the temperature. The results show significant increase of model quality due to optimal excitation signals used for the identification of the model parameters.

Paper C

J. UNGER, C. HAMETNER, S. JAKUBEK, M. QUASTHOFF: *A novel methodology for nonlinear system identification of battery cells used in non-road hybrid electric vehicles*. Journal of Power Sources, Volume 269, pages 883–897, Elsevier 2014.

In Paper C, the methodology of optimal model-based design of experiments and the local model network approach for battery modeling is presented including high current and temperature dependency. Results are shown for lithium-iron-phosphate as well as for lithium-polymer battery cells.

Paper D

J. UNGER, M. QUASTHOFF, S. JAKUBEK: *Innovative Energy Management System Using a Model Predictive Controller with Disturbance Prediction for Off-Road Applications*. 16. Antriebstechnisches Kolloquium (ATK 2015), 1. Auflage 2015, pages 427–443, 2015.

In Paper D, the concept of the EMS is presented in the context of non-road machinery and the improvement due to load and cycle prediction is discussed. The focus lies on the non-road machinery and the generic applicability of the methodology.

Paper E

J. UNGER, M. KOZEK, S. JAKUBEK: *Nonlinear model predictive energy management controller with load and cycle prediction for non-road HEV*. Control Engineering Practice, Volume 36, pages 120–132, Elsevier, March 2015.

In Paper E, the energy management system for a parallel hybrid electric powertrain is proposed, including the load and cycle prediction. The results measured at the test bed measurements show that a significant reduction in ICE dynamics is feasible, and fuel consumption as well as exhaust emissions can be reduced simultaneously by the proposed EMS. The results of Paper C are directly considered within the EMS as well as for the emulation of battery behavior.

Chapter 2
Battery Management

Abstract The battery management in non-road HEV is exposed to higher requirements compared to on-road HEV, since higher power densities and load dynamics
are usually demanded. For the control of HEV (see Chap. 4), a battery model comprising the nonlinear effects is required to be used in the controller itself as well as
essentially within the battery management system. The state of charge of the battery
is not measurable online, though, and needs to be estimated online during operation
(Hametner and Jakubek 2013). In this chapter, a generic methodology is proposed
comprising nonlinear system identification and optimal model-based design of experiments (DoE) of battery cells, which can be used for battery module modeling and
accurate SoC estimation during operation.

Keywords Lithium ion batteries · Nonlinear system identification · Optimal model
based design of experiments (DoE) · State observer

2.1 Introduction

2.1.1 Motivation

In many BMS, the open-circuit voltage (OCV) of the battery is used to estimate the
SoC, which is feasible as long as the battery is not in use. During operation, the
nonlinear behavior of the battery voltage comes into effect and big estimation errors
occur if only the OCV is used to estimate the SoC (Hu and Yurkovich 2012). The
integration of the battery current is another approach for SoC estimation, whose disadvantage is the drift due to the accumulation of current offsets when time increases.
Dynamic SoC estimators (e.g., extended Kalman filter) are a powerful way to estimate the SoC, but require a precise dynamic battery model for accurate estimation
(Plett 2004c). Precise battery models describe the nonlinear dynamic behavior of
the battery cell terminal voltage accurately by considering nonlinear battery effects
such as hysteresis, relaxation, and temperature effects. In general, due to the high

© The Author(s) 2016
J. Unger et al., *Energy Efficient Non-Road Hybrid Electric Vehicles*,
SpringerBriefs in Applied Sciences and Technology,
DOI 10.1007/978-3-319-29796-5_2

power densities in non-road HEV, the nonlinear effects of electrochemical batteries are increased (Gao et al. 2002), which complicates the modeling of the nonlinear battery effects (Unger et al. 2012a). Note that the used battery model needs to be real-time capable in order to be implemented in the BMS, which is in general a trade-off between accuracy and complexity.

2.1.2 Cell Chemistry-Dependent System Behavior of Batteries

The most known cell chemistry is the lead acid cell chemistry, which is used in almost every vehicle to start the engine. Beside of lead acid, there are many other cell chemistries such as lithium-iron-phosphate ($LiFePO_4$), lithium-polymer (LiPo), or nickel–metal hydride (Ni–MH). However, electrochemical batteries are strongly nonlinear systems, which depend nonlinearly on the SoC, temperature, and current, while additional effects such as relaxation and hysteresis are observable. In this context, relaxation refers to the slightly converging battery voltage at standby or steady-state current, while hysteresis refers to a phenomenon in relation to the cell polarization, which causes different shapes of the voltage values during charge/discharge of the battery. Inner chemical reactions may not be observed clearly in the voltage behavior since they occur randomly, but they are present and may have an influence on the voltage behavior. Lithium-ion batteries have a higher energy density compared to lead acid batteries and are therefore often used in mobile phones. Ni–MH chemistry has been used for traction batteries of HEV, but more and more traction batteries use the lithium-ion cells. This is not only caused by the higher energy density of lithium-ion cells but also by a smaller phenomenon referred to as *"memory effect,"* which reduces the capacity if the battery is not fully discharged before being charged again. A comparison of different battery cell chemistries in terms of energy and power capabilities is given in Fig. 2.1.

In general, traction batteries are built by coupling battery cells in serial and parallel connection in order to achieve a desired voltage level (series connection, since voltage add up) or battery capacity (parallel connection, since capacity add up). For the powertrain of non-road vehicles, the power capability of the traction battery is essential, due to which power-type battery cells are usually assembled in the battery module. Battery cells can be divided into power and energy cells, while power cells usually have higher power capabilities than the energy density maximized energy cells. The power capabilities of a battery cell can be expressed by the referred to as C-rate, which is a battery capacity-independent measure of the current intensity applied to a battery and is obtained by the quotient of current and battery cell capacity. In non-road vehicles, battery cells must be able to cope with possible C-rates above $20C$, while electric vehicles mostly require larger energy contents. Due to this reason, power cells are generally used in non-road HEV.

Furthermore, depending on the cell chemistry, voltage levels and battery behaviors are different, while even the shape of the discharge curve within the same cell chemistry may vary (e.g., lithium-iron-phosphate and lithium-polymer). Also rele-

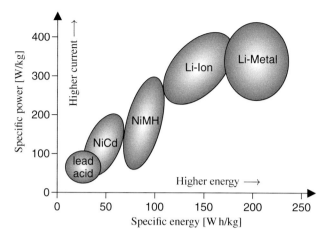

Fig. 2.1 Specific energy and power capabilities of different rechargeable battery cell chemistries

vant is the voltage level, which is significantly responsible for the overall energy flow and system requirements of the traction battery system. The type of the cell and the geometric structure (e.g., cylindrical, prismatic, …) mostly define the temperature behavior of the cell. This can be seen at higher C-rates, where the temperature increases significantly and limits the ability of the battery cells to be used in the high dynamic environment of non-road vehicles. In this context, the energetic efficiency plays a major role for the temperature behavior of the cells and is especially essential in non-road applications. Note here that the energetic efficiency is nonlinearly depending on the history of the battery cell's usage, which is important for the more than 10 years intended product life of a non-road vehicle. For the lithium-ion chemistry, additionally higher safety requirements are relevant, because overcharging with higher voltage may lead to explosion or fire and must be avoided in any case.

2.1.3 Challenges in Dynamic Battery Model Identification

In order to use a battery model approach within non-road vehicles, it must be generically applicable to any battery cell chemistry. A general model structure is therefore required that considers nonlinear effects. The nonlinear relations between the voltage and temperature as well as the SoC are unknown in advance, and only physically measurable variables are available for parameter identification. In order to identify specific nonlinear effects, long test runs must be performed because the reaction times of batteries are slow. Time consuming and expensive measurements are unacceptable for non-road applications, since the sales volume of non-road machines is low. Data-based approaches are methodologies which only depend on the provided data. The advantage of data-based models is that the model is flexible to any cell chemistry, while any model structure can be applied to consider the nonlinearities of the battery cells. On the other hand, the disadvantage is that an appropriate model structure needs to be found, and suitable data must be available for parameter identification. Nevertheless, an initial structure can be obtained from expert knowledge and included in the approach. The high dynamic requirement of non-road vehicles challenges especially the test equipment, because high current steps occur during operation. So far, high dynamics are not state of the art in the testing hardware of battery cells. Therefore, a major role corresponds to the design of experiments in order to achieve appropriate and reproducible measurements. The reproducibility of battery measurements depends on the excitation history of the battery, and appropriate procedures to clear the cell's short time history must be applied. Finding the appropriate procedure is challenging for battery cells, though, while the case is even more difficult for battery modules due to cell balancing. In order to gain maximum information from one test run, optimal design of experiments can be applied. Optimized test signals are able to achieve sufficient information content in the measurements, but since only the current is applied to the battery cell; in principle, a multi-dimensional optimization problem must be solved that consists of only one degree of freedom (DOF) and multiple effects to be tested.

2.1.4 State of the Art

The following sections review literature regarding battery cell modeling, design of experiments, battery module modeling, and SoC estimation.

2.1.4.1   Battery Cell Modeling

Three model approaches have been mainly used in the literature to model battery behavior:

1. Equivalent circuit models
2. Electrochemical battery models
3. Data-based battery models

In Fig. 2.2, the equivalent circuit model (ECM) approach is depicted schematically. Basic electric elements are used to describe the behavior of the terminal voltage, due to which physically interpretable parameters and real-time capability are achieved. Depending on the number of RC-circuits, different time constants are considered in the model. Only one RC-circuit accounting for nonlinear equilibrium potentials, rate and temperature dependencies, thermal effects, and response to transient power demand is used by Gao et al. (2002). A modified equivalent circuit model is used by Pattipati et al. (2010) to estimate SoC, state-of-health (SoH) and remaining useful life in the BMS. Due to the high power density appearing in the automotive industry,

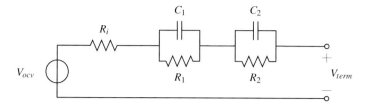

Fig. 2.2 Schematic overview of an equivalent circuit model (V_{term}) that consists of a constant voltage source (V_{ocv}), an inner resistance (R_i), and two RC-elements (R_1, C_1, R_2, C_2)

solution resistance, charge transfer resistance, and Warburg impedance cannot be neglected and must be considered within the ECM (Gomez et al. 2011).

Electrochemical battery models describe the electrochemical behavior of the battery by chemical reaction equations, which results in a physical model that is computationally intensive. The internal chemical states of the battery are simulated with high accuracy, and insight into the system is achieved (Klein et al. 2010), but in general the model is not real-time capable. Concentrated solution theory is used by Doyle et al. (1993) to describe lithium-ion battery cells. Klein et al. (2010) used partial differential-algebraic equations for state estimation, while a single particle model (SPM) is used by Santhanagopalan and White (2006) to estimate the SoC with an extended Kalman filter. However, the SPM model approach neglects the spatial variation of the states within the battery cell, which questions the validity for the operating region encountered for HEV (Chaturvedi et al. 2010).

Data-based system identification is a powerful approach for modeling and estimation purposes. Model structure and order are easily adaptable, although in general, physical interpretability is not given (Klein et al. 2010). In a series of three papers, Plett (2004a, b, c) proposed a SoC estimator based on a data-based nonlinear state-space model and an extended Kalman filter. The current direction is considered in the model, and hysteresis as well as relaxation is also included by a "hysteresis state" and a low-pass filter on the current, respectively. The model is assumed to be cell chemistry independent. Neural networks for battery modeling have been used by Charkhgard and Farrokhi (2010) and stochastic fuzzy neural networks by Wang (1994), Wang and Chen (2005), and Jing et al. (1998). A stochastic fuzzy neural network is also used by Wang et al. (2009) for the purpose of modeling the nonlinear dynamics of current, temperature, and SoC, while Xu et al. (2012) used it for the purpose of SoC estimation.

Hametner and Jakubek (2013) obtained a nonlinear battery model due to a local model network, which composes of several local models that are linear in their model parameters and have a certain area of validity defined by validity functions (see e.g., Hametner and Jakubek 2007; Murray-Smith and Johansen 1997; Gregorcic and Lightbody 2007). The nonlinear interpolation of the local linear models (LLM) achieves a nonlinear model output, while the model is constructed by an iterative algorithm. Starting with one global linear model, in each iteration, a LLM is added to the network until a certain threshold is reached (partitioning). Depending on the algorithm's strategy, the validity of the new LLM lies in a specific form in the partition space of the model. In order to identify the model parameters, any model approach requires measurements for parametrization. Battery cells are tested by applying a current excitation signal and recording the voltage response. Depending on the excitation signal, the testing device needs to fulfill certain requirements in terms of dynamics.

2.1.4.2 Design of Experiments

Simple constant discharge and charge cycles are used by Kroeze and Krein (2008) to identify the parameters of an ECM, while Gao et al. (2002), Chen and Rincon-Mora (2006), and Hentunen et al. (2011) used a discharge pulse excitation signal. Smith et al. (2010) also used a discharge pulse excitation signal for the model-based estimation of an electrochemical battery cell model. Charge and discharge modes within a pulse profile have been considered by Hu et al. (2011), Hu et al. (2009a), and Plett (2004b), in order to identify the parameters for more advanced ECMs (e.g., linear parameter-varying models). An asymmetrical current step profile, significantly more dynamic than the other excitation signals, has been used by Hu et al. (2009b) for the purpose of covering a wide range of SoC as well as a wide range of the current. The dynamic Federal Urban Driving Schedule (FUDS) is mostly used as a validation signal (see e.g., Smith and Wang 2006; Xu et al. 2012; Kroeze and Krein 2008; Sun et al. 2012a), but in non-road applications, the FUDS is rated as an example for low dynamics.

The design of experiments plays an important role, especially for data-based approaches, due to the decisive influence of the excitation signal on the parameter estimation (Unger et al. 2012a; Hametner and Jakubek 2013). In order to maximize the information content of measurements, model-based design of experiments can be used. The idea of model-based design of experiments is to use a prior process model (reference model) to maximize the information content of measurements in order to identify parameters with minimum variance (Pronzato 2008). A measure for the information content of an excitation signal can be obtained by the Fisher information matrix $\mathcal{I}$ (FIM), which gives the covariance for the parameters estimated from the excitation signal. Based on optimality criteria, the FIM is often used to optimize the excitation signal. Furthermore, depending on the accuracy of the reference model, constraints of the process can be considered in the excitation signal.

Static experiment design based on a local model network generation algorithm is proposed by Hartmann et al. (2011). Dynamic experiment design for multilayer perceptron networks is proposed by, e.g., Cohn (1996) and Deflorian and Klöpper (2009), where optimal inputs are chosen from a candidate set. Dynamic design of experiments based on multilayer perceptron networks is also centered by Stadlbauer et al. (2011a, b). These papers are used by Hametner et al. (2013b) to design non-linear dynamic experiments, which minimize the model variance of dynamic multilayer perceptron networks as well as local model networks. The influence of optimal model-based design of experiments on battery modeling, compared to dynamic excitation signals from the literature, is investigated by Unger et al. (2012a). Model-based design of experiments using a linear dynamic model and predefined current levels is proposed by Hametner and Jakubek (2013) in order to achieve optimal SoC excitation and minimal measurement duration.

2.1.4.3 Battery Module Modeling

The modeling of battery modules is different compared to battery cells, since battery balancing needs to be considered (Bentley 1997). Furthermore, the internal resistance of the battery cell connections as well as the temperature effect on the internal resistance of the battery plays major roles in the overall voltage behavior. Lee et al. (2010) provided a comprehensive review of joining technologies and processes for automotive lithium-ion battery manufacturing and discussed advantages and disadvantages of the different joining technologies, while corresponding manufacturing issues are mentioned. Since the battery forms a critical part of the HEV powertrain, Sen and Kar (2009) present a battery pack model that analyzes the variation of internal resistance as a function of temperature in order to provide the possibilities to design a cost-effective and efficient battery management system. Watrin et al. (2011) proposed a multiphysical battery pack model along with a test procedure that must be followed to obtain the different model parameters. Based on a screening process that provides a selection of battery cells with similar electrochemical characteristics, in Kim et al. (2011), the accuracy of the SoC estimator is increased. Due to the screening process and a parameter comparison, the battery module model can be simplified into a unit-cell model that is multiplied with the number of series connected cells. A similar capacity and resistance screening process is used by Kim et al. (2012) to improve the voltage/SoC balancing of a lithium-ion series battery module. Dubarry et al. (2008) use an equivalent circuit technique commonly applied for electrochemical impedance characterizations to describe the behavior of battery cells. These battery cell models are furthermore used to express the behavior of a battery module, while the imbalance of the battery cells is addressed along other effects to improve the battery module model. Battery module model accuracy can be increased significantly if intrinsic cell-to-cell variations in capacity and internal resistance are considered (Dubarry et al. 2009). An estimation of the remaining available power of a battery module is presented in Plett (2004d).

2.1.4.4 State of Charge Estimation

Many state of charge estimation algorithms are based on the open-circuit voltage (Lee et al. 2008). Since the relationship between the OCV and SoC is not identical for all batteries, Lee et al. (2008) proposed a modified OCV–SoC relationship to increase the SoC estimation accuracy. Commonly, an equivalent circuit model is used in combination with an extended Kalman filter (EKF) for the purpose of SoC estimation of batteries (see e.g., Do et al. 2009; Vasebi et al. 2007, 2008; Han et al. 2009; Bhangu et al. 2005), while Santhanagopalan and White (2006) use an electrochemical model. A sliding mode observer is applied by Kim (2006) to compensate the modeling errors of the used simple resistor–capacitor battery model. Neural networks (NNs) and EKF are used for modeling and SoC estimation by Charkhgard and Farrokhi (2010). Chen et al. (2011) used the same combination, but also developed a method to consider battery hysteresis effects. An adaptive unscented Kalman filtering method is proposed by Sun et al. (2011), which is further compared with an extended and an unscented Kalman filter. He et al. (2011) improved the dependence of the traditional filter algorithm on the battery model by an adaptive Kalman filter algorithm. An adaptive Luenberger observer for the SoC that uses an optimized model is built by Hu et al. (2010b). For the same purpose, a stochastic fuzzy neural network in combination with an extended Kalman filter for SoC estimation is proposed by Xu et al. (2012). Plett (2004b) based the SoC estimation also on an EKF, but used a state-space structure that considers the dynamic contributions due to open-circuit voltage, ohmic loss, and polarization time constants.

An alternative to the Kalman filter is the interacting multiple model (IMM) estimator, which has the ability to estimate the state of a dynamic system with several behavior models and to switch between them by corresponding rules (Mazor et al. 1998). The IMM is one of the most cost-effective hybrid state estimation schemes that can act as a self-adjusting variable-bandwidth filter, which is widely used for tracking maneuvering targets. Helm et al. (2012) used the IMM for a misfire detection that is based on two dedicated parametric Kalman filters. A fusion prediction-based interacting multiple model algorithms is used by Song et al. (2012). Another application of IMM is hypotheses merging (Blom and Bar-Shalom 1988). In the field of vehicle maneuvering, a fuzzy interacting multiple model unscented Kalman filter approach is presented by Jwo and Tseng (2009). Although the IMM approach could be applied for SoC estimation, so far, no papers are known in the literature discussing this topic.

Two different sliding mode observers for dynamic Takagi-Sugeno fuzzy systems are proposed by Bergsten et al. (2002). A nonlinear filter approach based on local linear models is proposed by Hametner and Jakubek (2013). Lendek et al. (2009) focused on the stability of cascaded fuzzy systems and observers.

2.1.5 Solution Approach

In this chapter, the methodologies are presented to achieve a precise battery cell terminal voltage model for non-road application, which can be applied to estimate the SoC of a battery module with high accuracy during operation.

The model is based on the data-based LMN approach, whose advantages are that expert knowledge can be considered, the computational effort is low, a random initialization of the parameters is avoided, and the LLMs can be interpreted as local linearization of the process (Hametner et al. 2012; Nelles 2001). Nelles and Isermann (1996) proposed the local linear model tree (LOLIMOT) construction algorithm, which is used to construct the LMN, while corresponding inputs are used to define the LMN structure. In order to consider the battery nonlinearities SoC, temperature and current as well as relaxation and hysteresis in the model, the inputs are adapted. In this context, a physically appropriate network is obtained by adapting the LOLIMOT algorithm to make use of a prepartitioned network and to prohibit splitting within specified dimensions of the network. The resulting battery cell model is applicable to different cell chemistries and real-time capable due to the low computational complexity.

Furthermore, optimal model-based design of experiments is utilized to achieve high dynamic excitation signals, which cover the entire SoC range during the measurements. Based on the Fisher information matrix $\mathcal{I}$, a scalar cost function $J(\mathcal{I})$ is used to optimize the excitation signal, while the battery cell is sufficiently excited and relaxation, hysteresis, as well as current and temperature effects are considered additionally. For this reason, the optimization is furthermore focused on real load ranges that are frequently used in operation (Unger et al. 2014). A gradient-based algorithm is used to solve the optimization problem, while battery constraints on current, voltage, and SoC are considered simultaneously. The obtained excitation signal reduces the identified model parameter variance and maximizes the information content of measurements.

Based on the optimal model-based DoE and the LMN approach, the battery module model can be built. A simple approach that multiplies the cell voltage with the number of cells in series connection is compared with the approach proposed by Kim et al. (2011), which considers the internal resistance of the battery cell connections as well as a voltage offset. The cell balancing is neglected, because in non-road vehicles, mostly a passive cell balancing strategy is used due to the appearing high energy conversions.

An accurate SoC estimation is then achieved for the battery module using the fuzzy observer approach as proposed by Hametner and Jakubek (2013). For a straight forward implementation, the battery module model must be transferred into a state-space representation with an augmented state vector that includes the SoC. At the end, different choices of process and measurement noise within the filter tuning show the trade-off between convergence and accuracy of the filter.

This chapter is organized as follows: First, the battery cell model is developed and an appropriate optimal model-based DoE is introduced for the developed battery cell model structure. Second, the temperature model approach is described. Third, different battery module model approaches that are based on the developed battery cell model are discussed. At the end, the SoC estimation for battery modules as well as for battery cells is discussed.

2.2   Data-Based Identification of Nonlinear Battery Cell Models

In this section, the generic methodology for nonlinear identification of high dynamic, current–voltage battery cell models is discussed. Nonlinear battery effects such as SoC, current, temperature, relaxation as well as hysteresis effects are considered by the model. First, the general architecture and structure of LMN is given, followed by the construction of the LMN using the LOLIMOT algorithm. Based on this, the final battery model is developed.

2.2.1   General Architecture and Structure of Local Model Networks

In principle, the local model network structure is built by local linear models that are only valid in a certain operation regime and interpolated to obtain the global nonlinear model output. An autoregressive with exogenous input model structure is chosen for the LLM, by what the regression vector φ follows with the global nonlinear model output $\hat{y}$, the global parameter vector θ, and the input variables u_l to

$$
\begin{aligned}
\varphi(k, \boldsymbol{\theta}) &= [\bar{\mathbf{y}} \ \bar{\mathbf{u}}_1 \ldots \bar{\mathbf{u}}_q \ 1]^T, \\
\bar{\mathbf{y}} &= \left[\hat{y}(k-1, \boldsymbol{\theta}) \ldots \hat{y}(k-n, \boldsymbol{\theta})\right], \\
\bar{\mathbf{u}}_l &= [u_l(k-d) \ldots u_l(k-d-m_l)], l = 1, \ldots, q,
\end{aligned} \tag{2.1}
$$

with the actual time instant k, the output order n, the input order m_l of the lth of q input variables, and the dead time d. Following Nelles and Isermann (1996), the input variables u_l span the so-called input space Q of the model. The bias is considered by the one in Eq. (2.1). Note that Eq. (2.1) is denoted for MISO systems, but MIMO systems can be modeled as well (Nelles 2001).

One challenge in data-based modeling is the optimal choice of the model order of the inputs and outputs. In D'Agostino (1986), different methodologies such as goodness-of-fit (GOF) techniques are recommended to find the optimal model order. GOF, from the statistical point of view, is discussed by Do Sá (2007). Akaike's information criterion trades off goodness-of-fit and model complexity and is used by, e.g., Yuan et al. (2013) for the optimal order of an ARX-structured battery model. As presented by Jackey et al. (2013), another methodology is to analyze different selections of the model order and choose the best compromise between complexity and accuracy. To this end, the mean squared error (MSE) provides a basis to make a decision.

The chosen model order is applied to the M LLMs, where the ith model output $\hat{y}_i$ is obtained by

$$
\hat{y}_i(k, \boldsymbol{\theta}) = \boldsymbol{\varphi}^T(k, \boldsymbol{\theta})\boldsymbol{\vartheta}_i, \tag{2.2}
$$

where $\boldsymbol{\theta} = [\boldsymbol{\vartheta}_1 \ldots \boldsymbol{\vartheta}_M]^T$ denotes the global parameter vector, which consists of the local parameter vectors $\boldsymbol{\vartheta}_i$ of all M LLM. Weighted aggregation of $\hat{y}_i$ leads to the global nonlinear LMN output

$$\hat{y}(k, \boldsymbol{\theta}) = \sum_{i=1}^{M} \hat{y}_i(k, \boldsymbol{\theta}) \Phi_i(k), \tag{2.3}$$

where $\Phi_i(k)$ is the validity function of the ith LLM. Using a Kernel function $\mu_i(k, \mathbf{z})$, which can be any common function (e.g., uniform, triangle, ...) (Nelles 2001), the validity function follows by

$$\Phi_i(k) = \frac{\mu_i(k, \mathbf{z})}{\displaystyle\sum_{j=1}^{M} \mu_j(k, \mathbf{z})}, \tag{2.4}$$

where $\mathbf{z} = \begin{bmatrix} z_1 \ldots z_\phi \end{bmatrix}$ span the so-called partition space $\mathcal{Z}$ of the model using the ϕ partition variables (Gregorcic and Lightbody 2007). Note that the normalization results from the sum of all validity functions, which is required to be 1. Although input and partition variables may be used in the partition space $\mathcal{Z}$ and input space $\mathcal{Q}$ simultaneously, they are not mandatorily the same (Hoffmann and Nelles 2001).

2.2.2 Construction of LMN Using LOLIMOT

A LMN is usually constructed by algorithms, which iteratively add new LLM to the network. One such algorithm is the local linear model tree algorithm presented by Nelles (2001). LOLIMOT searches for the worst LLM by the quadratic error criterion and identifies the dimension in which a split of the model could achieve the best improvement for the global model output. The worst LLM is then axis-orthogonally split into two new models in the dimension with the best improvement (Nelles and Isermann 1996). Iteratively, the LMN grows until a certain threshold (here the maximal number of LLM M) is reached.

The parameters for the two new models are determined by weighted least squares (WLS). In case of numerical effects due to significant differences in the input, partition, and output variables, all signals used within the algorithm are normalized from 0 to 1 (Nelles 2001). Schematically, the LOLIMOT procedure for a two-dimensional partition space is depicted in Fig. 2.3. In every iteration, the quadratic evaluation criterion is used to find the worst LLM. The worst model is split into all possible dimensions, while the best alternative, identified likewise with the quadratic evaluation criterion, is chosen.

A characteristic of the LOLIMOT algorithm is the Kernel function $\mu_i(k, \mathbf{z})$, which is chosen to be Gaussian (Nelles 2001):

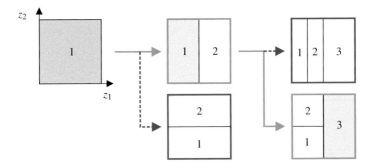

Fig. 2.3 Decision process of the LOLIMOT algorithm in a two-dimensional partition space

$$\mu_i(k, \mathbf{z}) = \exp\left(-\frac{1}{2}\left(\frac{(z_1(k) - c_{i1})^2}{\sigma_{i1}^2} + \cdots + \frac{\left(z_\phi(k) - c_{i\phi}\right)^2}{\sigma_{i\phi}^2}\right)\right). \qquad (2.5)$$

Since the partition space $\mathcal{Z}$ is split axis-orthogonally, c_{ij} denotes the center point of the LLM and σ_{ij} is the corresponding individual standard deviation, which is approximated by

$$\sigma_{ij} = k_{\sigma,j}\Delta_{ij}. \qquad (2.6)$$

The term Δ_{ij} corresponds to the spread of the LLM, while $k_{\sigma,j}$ is a user-defined sharpness factor that can be seen as an LLM overlapping factor that influences the smoothness of the nonlinear model output. Note that the optimal sharpness factor varies depending on the specific application as well as the partitioning dimensions.

Alternative construction algorithms have the same aim, but are different to LOLIMOT. Jakubek and Keuth (2006) used statistical criteria along with regularization to allow an arbitrary orientation and extent in the partition space of the constructed LMN. In Jakubek and Hametner (2009), a proper partitioning of the LMN is achieved by an expectation–maximization algorithm that makes use of a residual obtained from generalized total least squares parameter estimation. The advantage of LOLIMOT compared to mentioned alternatives is the low implementation complexity due to the axis-orthogonal orientation, for which reason LOLIMOT is used in this book.

2.2.3 Battery Cell Modeling Using LMN

So far, the methodology for battery modeling has been discussed. In the following, the procedure to apply the LMN to battery modeling is presented. Corresponding inputs and the structure for the LMN are presented in order to consider nonlinear battery effects and enhancements for the LOLIMOT algorithm to increase the physical

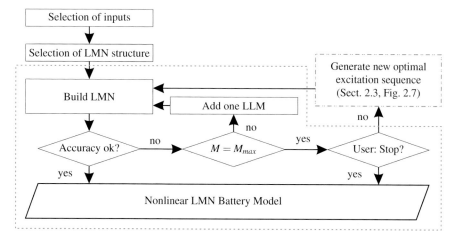

Fig. 2.4 Construction process flowchart for the battery model identification

meaning of the battery model. Figure 2.4 depicts an overview of the battery model construction process using a flowchart.

2.2.3.1 Corresponding LMN Inputs of Nonlinear Battery Cell Effects

Electrochemical batteries comprise physical and chemical nonlinear effects, which can be observed and explained in detail on the basis of Figs. 2.5 and 2.6.

Figure 2.5 depicts constant charge and discharge curves for different C-rates over the SoC. Cell **A** refers to a lithium-polymer chemistry, while Cell **B** represents a

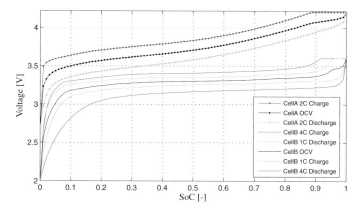

Fig. 2.5 Open-circuit voltages and different constant current discharge/charge curves from a lithium-polymer (Cell **A**) and a lithium-iron-phosphate (Cell **B**) cell

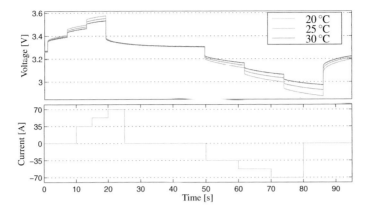

Fig. 2.6 Measurements of voltage responses obtained from a lithium-iron-phosphate cell (Cell *B*) during an applied current step sequence at different temperatures

lithium-iron-phosphate chemistry. In order to be able to plot the voltage curves over the SoC, the cell current is integrated (Sun et al. 2012a), while interpolation between the charge and discharge curve provided an estimate for the open-circuit voltage (Abu-Sharkh and Doerffel 2004).

In Fig. 2.6, a current step sequence at different temperatures measured for Cell *B* is shown. The nonlinear physical influence of current and temperature can be observed, but distinction must be made between physical and chemical effects. Physical effects are caused by any physical interaction such as an applied current and the resulting temperature increase, while chemical effects arise also without physical interaction such as hysteresis and relaxation. A dynamic change in the battery voltage is caused by the applied battery cell current, while the voltage drop is directly influenced by the temperature. Due to the lower/higher temperature, increased/decreased internal resistance of the battery cell is obtained and a bigger/smaller voltage drop is caused. Current as well as temperature are physically measurable online and can directly be considered in the model. Hence, the corresponding inputs $u_{Current}$ and z_{Temp} are selected. The time constant of the temperature is significantly higher than the time constant of the current. Due to this reason, the current is included in the input space $\mathcal{Q}$, while the temperature is assumed to be static and is therefore included in the partition space $\mathcal{Z}$. Note that the small temperature gradients compared to gradients of the current underline the assumption.

The nonlinear influence of the SoC on the battery cell voltage is clearly shown in Fig. 2.5. Since the SoC cannot change dynamically or independently to the current, the SoC is considered as corresponding static input z_{SoC} with the value of the actual SoC. The value of the SoC is not measurable online and needs to be estimated in real batteries (see e.g., Hametner and Jakubek 2013; Plett 2004c). For simulation purposes, the current can be integrated to determine a value for the SoC. Note that in Sect. 2.6, the SoC estimation is discussed in detail.

Chemical effects are not directly measurable and need to be provided indirectly by corresponding inputs. The slightly to a steady-state value converging voltage at standby current in Fig. 2.6 is referred to as relaxation (Plett 2004b), although relaxation also acts during current phases (Bernardi and Go 2011). In Plett (2004b), a low-pass filter on the current, which follows certain requirements, is used to model the relaxation, whose time constant is significantly different to the one of the currents. Following Plett (2004b),

- after a long rest period and
- during constant current discharge/charge,

the relaxation state needs to converge to zero, which can be realized by the dynamic corresponding input

$$u_{\text{Relax}} = filt(\Delta u_{\text{Current}}). \tag{2.7}$$

Note that in order to force the filter $filt(\cdot)$ to have zero DC-gain, the change rate of the current $\Delta u_{\text{Current}}$ is used. For lithium-iron-phosphate as well as lithium-polymer chemistry, a third-order low-pass filter showed to be appropriate, whose relaxation time constant is approximated properly on the basis of the voltage converging speed at standby current after a current pulse has been applied to the battery cell (Unger et al. 2014).

Different shapes of the charge and discharge curve in Fig. 2.5 indicate the nonlinear chemical effect referred to as hysteresis (Tang et al. 2008). At standby current in Fig. 2.6, the effect is observable as well. In principle, the hysteresis effect separates the model into a charge and discharge model, which can be achieved using $sign(\cdot)$ of the current as a corresponding first-order hysteresis input z_{Hyst} that is kept at the last value, if the current is zero (Unger et al. 2014). As hysteresis is acting statically on the battery behavior, z_{Hyst} is included in $\mathcal{Z}$. Intercalation effects, conductivity of anode/electrolyte/cathode, concentration gradients, or other known chemical effects play a tangential role compared to the mentioned effects and are therefore neglected.

2.2.3.2   LMN Structure of Battery Cell Models

Distinction must further be made between dynamic and static influence on the cell voltage. Dynamic influence of the inputs needs to be considered in the dynamic LLM and must therefore be included in the input space $\mathcal{Q}$, while static influence is important for the partitioning and must therefore be included in the partition space $\mathcal{Z}$. This defines the structure of the LMN and leads to

$$\begin{aligned} \mathcal{Z} &= [z_1 \ z_2 \ z_3] \ \widehat{=} \ \left[z_{\text{SoC}} \ z_{\text{Hyst}} \ z_{\text{Temp}} \right], \\ \mathcal{Q} &= [u_1 \ u_2 \ u_3] \ \widehat{=} \ \left[z_{\text{SoC}} \ u_{\text{Current}} \ u_{\text{Relax}} \right]. \end{aligned} \tag{2.8}$$

To this end, the inclusion of the SoC z_{SoC} in the input space $\mathcal{Q}$ implies two advantages. On the one hand side, the continuous change of the voltage depending on the SoC is considered and on the other hand, the observability of the SoC within the model is given, which is required by the SoC estimation with a fuzzy observer (c.f. Hametner and Jakubek 2013).

2.2.3.3   User-Defined Prepartitioning of the LMN Structure

In the iterative construction of the LMN, only the quadratic evaluation criterion decides whether to split a model or not. This may lead to physically inappropriate partitions that make a physical interpretation impossible. Due to this reason, LOLIMOT is enhanced to use initial partitions of $\mathcal{Z}$ instead of one global partition. The predefined physically appropriate partitions reduce the computational efforts, while all partitions can be kept physically appropriate if selected dimensions of the partition space are prohibited to be split (Unger et al. 2014). This influences the partitioning significantly and needs to be discussed in detail in the following.

Prohibiting a dimension to be split is only appropriate if a physical reason is given to limit the number of LLM within this dimension. At this point, expert knowledge can be included to improve the model accuracy. The hysteresis input z_{Hyst} refers to the corresponding cell polarization, which can only reach two different states. In this case, an initial split into charge as well as discharge mode and prohibiting to split within this dimension is advantageous. From Fig. 2.6, the temperature influence seems to be evenly distributed, and therefore three initial partitions are defined for the temperature input. Note that due to the split in the hysteresis dimension, a simultaneous split of charge and discharge mode cannot be achieved by definition, because of which partitioning is prohibited within the temperature dimension. The number of initial partitions follows to 6 and the partitioning DOF is limited to the SoC dimension.

2.3   Optimal Model-Based Design of Experiments

This section discusses the optimal model-based design of experiments for battery model identification. The goal of optimal model-based DoE is to obtain an excitation signal that minimizes the variance of the identified model parameters. To this end, the system dynamics of a battery cell must be sufficiently excited, while the entire SoC range is covered and relaxation as well as hysteresis effects are considered. The battery system behavior is obtained by applying a load current excitation signal $\mathbf{U}$ to the battery cell and recording the cell terminal voltage. In on-road applications, intermediate current steps show sufficient dynamics (Hu et al. 2009b), but in non-road applications higher dynamic excitation signals are required (Unger et al. 2012a).

For that purpose, a methodology is proposed to obtain optimal excitation signals for non-road application. In the following, first, an a priori available battery model and the Fisher information matrix $\mathcal{I}$ are used to formulate optimality criteria. Second, a constrained optimization problem is formulated to optimize an excitation signal under consideration of constraints and the nonlinear effects of a battery cell. Note that the information content is further improved by focusing the optimization especially on load ranges frequently used in operation. Third, the optimization by means of a gradient-based algorithm is described in detail. At the end, some extensions are made to the obtained optimal excitation sequence to take into account the entire SoC operating range, relaxation, hysteresis, and constant current behavior.

2.3.1 Optimization Criteria Based on the Fisher Information Matrix

The Fisher information matrix $\mathcal{I}$ is a tool to measure the information content of measurements in terms of the covariance of the estimated model parameters. In order to increase the information content, the system inputs by means of the excitation signal $\mathbf{U}$ need to be modified (Hametner et al. 2013b). The calculation of $\mathcal{I}$ is based on the parameter sensitivity vector $\boldsymbol{\psi}(k)$, which is the partial derivative of the model output with respect to the model parameters. For the LMN approach as described in Sect. 2.2, $\boldsymbol{\psi}(k)$ follows by

$$\boldsymbol{\psi}(k) = \frac{\partial \hat{y}(k, \boldsymbol{\theta})}{\partial \boldsymbol{\theta}} = \begin{bmatrix} \Phi_1(k)\boldsymbol{\varphi}(k, \boldsymbol{\theta}) \\ \vdots \\ \Phi_M(k)\boldsymbol{\varphi}(k, \boldsymbol{\theta}) \end{bmatrix}, \quad k = 1, \ldots, N, \tag{2.9}$$

where M denotes the number of LLM and $\Phi_i(k)$, $\boldsymbol{\varphi}(k, \boldsymbol{\theta})$, $\boldsymbol{\theta}$ as defined in Eqs. (2.4) and (2.1). A reference model is required for the optimization and can be obtained by two possibilities (Unger et al. 2014):

- A LMN model is available.
- A different model (no LMN model) or measurements are available.

A LMN with only one LLM describes a linear model of the battery and can be identified easily based on a priori available measurements. In this book, the easy approach of only a linear model is followed to show the significant influence of the methodology to the model quality. Note that a reference LMN can be identified by simulation data created by an available complex electrochemical model, alternatively.

The Fisher information matrix is defined by

$$\mathcal{I} = \frac{1}{\sigma^2} \sum_{k=1}^{N} \underbrace{\frac{\partial \hat{y}(k, \boldsymbol{\theta})}{\partial \boldsymbol{\theta}} \frac{\partial \hat{y}(k, \boldsymbol{\theta})}{\partial \boldsymbol{\theta}}^{T}}_{\boldsymbol{\psi}(k)}, \quad k = 1, \ldots, N, \tag{2.10}$$

where σ is the variance of the measurement noise (Goodwin and Payne 1977). Denoting $\boldsymbol{\Psi}$ as

$$\boldsymbol{\Psi} = \left[\boldsymbol{\psi}^T(1) \ldots \boldsymbol{\psi}^T(N) \right]^T, \tag{2.11}$$

the FIM can be expressed by

$$\mathcal{I} = \frac{1}{\sigma^2} \boldsymbol{\Psi}^T \boldsymbol{\Psi}. \tag{2.12}$$

Based on the obtained FIM, in the literature three common scalar criteria are known for the optimization of the excitation sequence. They are formulated as follows (Goodwin and Payne 1977):

$$\text{A-optimality: } J_A = \text{Tr}\left(\mathcal{I}^{-1}\right) \to \min_{\mathbf{U}} \tag{2.13}$$

$$\text{D-optimality: } J_D = \det\left(\mathcal{I}\right) \to \max_{\mathbf{U}} \tag{2.14}$$

$$\text{E-optimality: } J_E = \lambda_{min}\left(\mathcal{I}\right) \to \max_{\mathbf{U}} \tag{2.15}$$

A-optimality minimizes the trace of the inverse of the Fisher information matrix, D-optimality corresponds to the maximization of the determinant of the FIM, and E-optimality is targeted to maximize the smallest eigenvalue of the FIM. The advantage of D-optimality is the higher sensitivity to single-parameter covariances compared to the A-optimality (Stadlbauer et al. 2011a).

Following the three common criteria Eqs. (2.13)–(2.15), Fig. 2.7 shows the process to obtain the optimal excitation signal **U**.

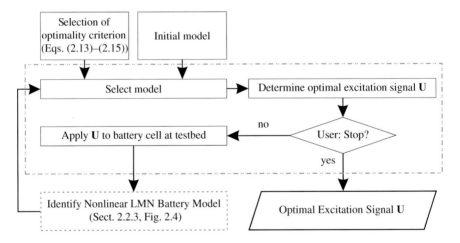

Fig. 2.7 Flowchart of the optimal design of experiments for battery models. The *dot* and *dash line* refer to the LMN identification as depicted in Fig. 2.4

2.3.2 Formulation of the Constrained Optimization Problem

The optimization aims primarily to achieve sufficient high dynamic currents within the excitation sequence, while constraints on current and battery cell voltage are considered. Note that the SoC is limited due to the physical capacity of the battery, which is automatically considered by voltage and current constraints. Constraints must be kept to avoid, among other things, physical damage, accelerated life-time reduction, electrolyte oxidation, fire, or explosion. Especially, the lithium-ion chemistry is sensitive to overcharge and over-voltage, respectively, which attracts the attention due to safety issues.

The current is furthermore constraint to ranges frequently used in operation to increase the information content especially in ranges used in real applications. Consequentially, two possibilities can be used for current constraints:

1. High dynamic excitation between physical minimum/maximum current.
2. High dynamic excitation between load ranges, frequently used in operation.

The first approach simply defines the current constraints at the physical minimum and maximum values, while the second approach defines the constraints using a real load cycle analysis that is realized as follows:

step 1. Select a representative load cycle.
step 2. Determine the distribution density of the load by a histogram with a defined number of intervals.
step 3. Set the lower and the upper current constraints corresponding to the interval limits of the histogram.
step 4. Define the durations within the corresponding constraint ranges by the corresponding distribution densities.
step 5. Scale the constraints corresponding to the SoC.

Figures 2.8 and 2.9 show the load cycle analysis in more detail: A scaled real load cycle, for which the analysis is done, can be seen in the left subplot of Fig. 2.8. In the middle subplot, the corresponding histogram is depicted, while to the right, the current constraints for the depicted load cycle are shown.

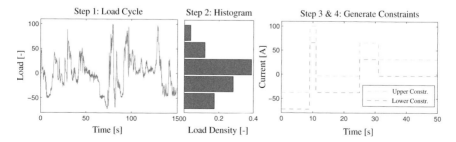

Fig. 2.8 Step 1–4 in the constraint construction used for the optimization of the excitation signal

Fig. 2.9 Step 5 in the
constraints construction:
SoC dependent limitation of
the minimal/maximal battery
current

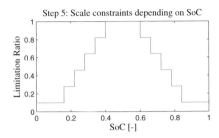

In order to consider output constraints within the experiment design, the reference
model must have sufficient accuracy. The used linear reference model is not able to
provide such a precision, though, which implicates that voltage constraints must be
included indirectly through the current. To this end, a limitation ratio for the current
depending on the SoC is introduced. Figure 2.9 depicts the limitation ratio. Note that
the reliability of the output constraints in general depends on the model accuracy and
increases with each model update (Hametner et al. 2013a). Nevertheless, due to the
maximal current constraints, the maximal deviation of the SoC from the starting SoC
value is limited (Unger et al. 2014). The SoC limits therefore follow by an adequately
chosen starting SoC value.

On the basis of these constraints, a formulation of the optimization problem can be
obtained. Following Pronzato (2008), the D-optimality criterion has higher sensitivity
to single-parameter covariances and is more invariable to re-parametrization of the
model than the A-optimality. Due to this reason, the D-optimality criterion is used
to formulate the optimization problem:

$$\text{D-optimality:} \quad \max_{\mathbf{U}} \ \det\left(\mathcal{I}\right) \tag{2.16}$$

$$\text{s.t.} \quad \begin{cases} U_{min}(k) \leq U(k) \leq U_{max}(k), & k = 1, \ldots, N \\ \mathbf{U} \in \mathbb{R}^{N \times 1} \end{cases},$$

where $U = u_{\text{Current}}$ corresponds to the current input. Note that relaxation input and
SoC are directly dependent on the current, which is therefore the only degree of
freedom within the optimization.

2.3.3 Constrained Optimization

The constraints require to solve the stated optimization problem iteratively. One
approach with low computational efforts and sufficient performance is the gradient
descent method. The derivative of the design criterion with respect to the input $U(r)$
for all observations N composes the gradient $\mathbf{g} = [g(1) \ldots g(N)]^{T}$. In order to obtain
the single gradients, the trace of the product of the derivative of the determinant of

the FIM with respect to the parameter sensitivity matrix $\mathbf{\Psi}$ and the derivative of $\mathbf{\Psi}$ with respect to $U(r)$ need to be built, which is computationally intensive (Stadlbauer et al. 2011a). Nevertheless, the rth observation follows by

$$\frac{d J_D(\mathbf{\Psi})}{d U(r)} = \mathrm{Tr}\left(\frac{d J_D(\mathbf{\Psi})}{d \mathbf{\Psi}^T}\frac{d \mathbf{\Psi}}{d U(r)}\right), \tag{2.17}$$

where for the D-optimality, the first term is obtained by (cf. Magnus and Neudecker 1988)

$$\frac{d J_D(\mathbf{\Psi})}{d \mathbf{\Psi}^T} = 2 J_D(\mathbf{\Psi}) \mathbf{\Psi} [\mathbf{\Psi}^T \mathbf{\Psi}]^{-1}. \tag{2.18}$$

Note that the inversion of the FIM appears in Eq. (2.18), due to which the FIM is required to be a regular matrix with full rank (Hametner et al. 2013b).

The second term in Eq. (2.17) is obtained by the single derivatives of the parameter sensitivity vectors with respect to the model input, which are based on the derivative of the regressor $\varphi(k, \boldsymbol{\theta})$ as defined in Eq. (2.1) with respect to the input $U(r)$. Denoting the derivative of $\varphi(k, \boldsymbol{\theta})$ by

$$\frac{d \varphi^T(k, \boldsymbol{\theta})}{d U(r)} = \left[\frac{d \hat{y}(k-1, \boldsymbol{\theta})}{d U(r)} \cdots \frac{d \hat{y}(k-n, \boldsymbol{\theta})}{d U(r)}\quad \delta_{1l}\delta_{(k-1)r} \ldots \delta_{pl}\delta_{(k-m_l)r}\; 0\right], \quad k > r, \tag{2.19}$$

where δ_{ij} is the Kronecker delta function and l corresponds to the current input of the LMN battery model. The former model output with respect to the input $U(r)$ is recursively calculated and follows to

$$\frac{d \hat{y}(k, \boldsymbol{\theta})}{d U(r)} = \frac{\partial \hat{y}(k, \boldsymbol{\theta})}{\partial \hat{y}(k-1, \boldsymbol{\theta})} \cdot \underbrace{\frac{d \hat{y}(k-1, \boldsymbol{\theta})}{d U(r)}}_{\text{recursive calculation}}$$

$$+ \frac{\partial \hat{y}(k, \boldsymbol{\theta})}{\partial \hat{y}(k-n, \boldsymbol{\theta})}\frac{d \hat{y}(k-n, \boldsymbol{\theta})}{d U(r)} + \frac{\partial \hat{y}(k, \boldsymbol{\theta})}{\partial U(r)}, \quad k > r. \tag{2.20}$$

Based on Eqs. (2.19) and (2.20), the derivative of the parameter sensitivity vector with respect to the model input is obtained by

$$\frac{d \boldsymbol{\psi}(k)}{d U(r)} = \begin{bmatrix} \Phi_1(k)\dfrac{d \varphi(k, \boldsymbol{\theta})}{d U(r)} \\ \vdots \\ \Phi_M(k)\dfrac{d \varphi(k, \boldsymbol{\theta})}{d U(r)} \end{bmatrix} + \begin{bmatrix} \varphi(k, \boldsymbol{\theta})\dfrac{d \Phi_1(k)}{d U(r)} \\ \vdots \\ \varphi(k, \boldsymbol{\theta})\dfrac{d \Phi_M(k)}{d U(r)} \end{bmatrix}, \quad k > r, \tag{2.21}$$

while the compact notation of Eq. (2.21) yields the second term of Eq. (2.17)

$$\frac{d\boldsymbol{\Psi}}{dU(r)} = \left[\frac{d\boldsymbol{\psi}^T(1)}{dU(r)} \cdots \frac{d\boldsymbol{\psi}^T(N)}{dU(r)}\right]^T.$$ (2.22)

Using Eqs. (2.18) and (2.22), the gradient $\mathbf{g}$ can be used in the gradient descent method to update the excitation sequence. Considering the current constraints along with an adaptively adjusted step size η, the update follows by

$$\mathbf{U}^{(\nu+1)} = \mathbf{U}^{(\nu)} + \eta \cdot \mathbf{g}^{(\nu)}, \quad \nu = 0, 1, 2, \ldots$$

$$\text{s.t.} \begin{cases} \mathbf{U}_{min} \leq \mathbf{U}^{(\nu+1)} \leq \mathbf{U}_{max} \\ \mathbf{U} \in \mathbb{R}^{N \times 1} \end{cases},$$ (2.23)

which process is repeated until no further improvement is achieved.

Note that the voltage constraints are included indirectly, and therefore a nonlinear constrained optimization problem is avoided. Nevertheless, several alternative approaches are known, which include the output constraints directly in the formulation of the optimization problem: The (quadratic) difference between the gradient of the design criterion and the excitation signal increment is minimized by Stadlbauer et al. (2011a), while the feasible area is approached simultaneously. In order to realize the constrained optimization, Hametner et al. (2013b) applied Lagrangian multipliers. Sequential quadratic programming (see e.g., Luenberger and Ye 2008) and numerical multi-objective optimization (see e.g., Seyr and Jakubek 2007) are further approaches.

2.3.4 Extensions on the Excitation Sequence

The previous sections discussed the optimization of the excitation sequence, while constraints are considered. For the purpose of especially including the battery-specific nonlinear effects, the optimized excitation sequence is extended in the following.

Constant discharge/charge currents with following standby currents are able to reveal the voltage behavior caused by relaxation and to discharge/charge the battery to a desired SoC value. Constant discharge/charge current pulses with subsequent standby current are therefore added in front of/after the optimized sequence. Following the limitation ratio (see Fig. 2.9), the values of the constant currents are obtained, while the duration of the pulses is related to the longest time constant of the system. Sufficient information about relaxation requires the duration of the standby current to be at least a multiple of the duration of the constant current. An analysis of Fig. 2.6 leads to a duration ratio of at least 6 for an abated voltage degradation at standby current. Due to this reason, a duration ratio of 8 is suggested and used in this book.

The final excitation signal is obtained by merging the extended high dynamic sequences for evenly distributed SoC values across the entire SoC range of interest (Unger et al. 2014). Any SoC deviations from the desired SoC are carefully compensated by varying the durations of the constant currents, in order to achieve information across the entire SoC range. Note that the SoC range of interest for different non-road applications differs from each other and therefore may vary.

2.4 Temperature Model of Battery Cells

In order to provide a simulation model, the battery temperature needs to be modeled as well. At high charge and discharge currents, the cell temperature increases significantly and may raise above allowable limits (Onda et al. 2006). Hu et al. (2010a) proposed an accurate battery thermal model using a Foster network, which extracts capacitance and resistance from computational fluid dynamics (CFD) results. A simplified mathematical model is presented by Jeon and Baek (2011) that considers heat generations due to joule heating and entropy change, respectively. Sun et al. (2012b) developed a three-dimensional thermal model to gain a better understanding of the thermal battery cell behavior in a battery module. The more advanced approach considers the battery nonuniform heat generation rate, the battery temperature distribution as well as battery temperature variation across the module in order to predict the temperature behavior during simulated driving cycles. A lumped-parameter thermal model based on a differential equation is developed in Forgez et al. (2010), whose approach is used in this book. Based on the cell thermal capacity c_p and the heat transfer coefficient h_{out}, the temperature behavior of a battery cell can be expressed by

$$c_p \frac{d\vartheta_{cell}}{dt} = h_{out} \left(\vartheta_{amb} - \vartheta_{cell} \right) + I_{cell}^2 R_{int}, \tag{2.24}$$

where ϑ_{amb} is the ambient temperature, ϑ_{cell} is the cell temperature, R_{int} is the internal resistance of the cell model, and I_{cell} is the cell current. The battery type (energy or power cell) influences the temperature behavior essentially, since the losses produce heat, which is dissipated through the battery cell. In general, energy cells have a different design, due to which the heat transfer resistance is higher. Nevertheless, as can be seen in Eq. (2.24), the internal resistance R_{int} is required and needs to be exported from the LMN battery cell. Due to the nonlinear behavior of the battery cell, R_{int} depends on the actual state of the battery cell, though.

In this context, the interpretability of the LMN battery model should be discussed in more detail. Due to the physically appropriate partitioning of the LMN, the steady-state gain of the LLM can be interpreted as the local internal resistance of the particular LLM. Note that the calculated steady-state gain is normalized and needs to be transformed first before it can be interpreted as physical value. In Fig. 2.10 in the upper two subplots, the identified internal resistances for Cell *A* and Cell *B*, respectively, are shown for different temperatures. Since the partition space is three-dimensional,

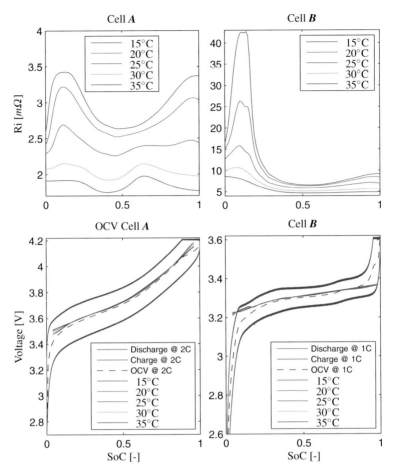

Fig. 2.10 Interpretability of LMN battery model parameters shown at Cell **A** and Cell **B**. The current dimension is held at $I_{cell} = -40\ A$

the current is held at $I_{cell} = -40\ A$ to be able to depict a two-dimensional representation of the parameters. The lower two subplots show the identified open-circuit voltages in comparison with an OCV that is obtained from an interpolation of the charge and discharge curves at 2C. As can be seen, the OCV is precisely identified, though the excitation signal has been highly dynamic and included only few standby phases during the measurements. Based on this interpretation, look-up tables for the internal resistance R_{int} and the OCV V_0 can be exported from the LMN and used in Eq. (2.24).

2.5 Battery Module Model Design

Battery module models are required for the simulation of powertrain concepts in hybrid electric vehicles as well as for the implementation in battery simulators, which emulate the real battery behavior at the test bed. Real-time capability is especially necessary for the implementation in battery simulators. Nevertheless, battery modules consist of multiple battery cells connected in series and parallel connections, which results in a required monitoring system that observes the voltage and SoC balance between the battery cells to avoid physical damage. The balancing strategy needs to be included in a battery model, as long as there is an effect during operation. Due to this reason, cell balancing in battery modules is reviewed in the following, before the design of the LMN-based battery module model is discussed.

2.5.1 Battery Cell Balancing in Battery Modules

Battery balancing refers to the equalization of the state of charge of the battery cells within a battery module to avoid overcharge, which might cause physical damage and safety issues due to explosion and fire. Cao et al. (2008) present the theory behind balancing methods for battery systems within the past 20 years and group their nature of balancing. In general, two different balancing techniques are known (Andrea 2010):

1. Active balancing
2. Passive balancing

Active balancing draws the energy from the most charged cell and transfers the energy through DC–DC converters to the least charged cells. For example, Lee and Cheng (2005) proposed an active balancing algorithm for lithium battery modules. Passive systems waste energy from the most charged cell as heat, until the cell charges are equalized. To this end, the balancing strategy needs to be considered in the battery module model, if cell balancing is carried out during operation and has an influence on the voltage behavior.

In this book, a battery module with implemented passive balancing system is used. Thus, balancing is neglected in the proposed model approach. Nevertheless, different balancing strategies and their implementation in battery models are discussed in, e.g., Bentley (1997), Daowd et al. (2011), and Weicker (2013).

2.5.2 LMN-Based Battery Module Design

In non-road vehicles, the voltage level of battery modules is usually high, because the current capabilities of electrochemical batteries for high power demands are

generally limited. Thus often super capacitors are used due to their significantly
larger power capabilities. Nevertheless, the higher energy densities of batteries make
them an equal alternative. During the design process of hybrid electric powertrains,
the planned battery module may not be physically available. This complicates the
design of battery module models, since a sufficient parametrization is difficult. To this
end, four different approaches of battery module model approaches Σ_i are discussed
in the following in order to show the influence of different considered effects.

Battery cell measurements are easier to realize, and therefore a battery module
model that is based on battery cell models is desired and advantageous. The proposed methodology in the previous sections provides LMN battery cell models that
can easily be obtained by cell measurements, but does not consider the additional
internal resistance of battery modules resulting from the cell connections. Nevertheless, the internal resistance is unknown in advance, and therefore a first obvious
model approach is

$$V_{module,\Sigma_1} = n_{cells} \cdot V_{cell}, \tag{2.25}$$

where n_{cells} is the number of series connected cells and V_{cell} refers to the battery
cell voltage. The battery cell voltage V_{cell} can be obtained by the LMN battery cell
model using the battery cell current

$$I_{cell} = \frac{I_{module}}{p_{cells}}, \tag{2.26}$$

where p_{cells} is the number of parallel connected cells, and I_{module} is the battery
module current. Note that the internal resistances of the battery cell connections
are not considered in this simple approach and may limit the accuracy. Therefore,
however, no information is required about the internal resistance of a battery module.

Considering the internal resistance of the battery module in V_{module,Σ_1} leads to
model Σ_2

$$V_{module,\Sigma_2} = n_{cells} \cdot V_{cell} - R_{int,\Sigma_2} \cdot I_{module}, \tag{2.27}$$

where R_{int,Σ_2} refers to the internal module resistance. Battery cells usually vary in
nominal cell voltage, which can be considered by an offset voltage. Module model
approach Σ_3 follows therefore to

$$V_{module,\Sigma_3} = n_{cells} \cdot V_{cell} - R_{int,\Sigma_3} \cdot I_{module} + V_{offset,\Sigma_3}, \tag{2.28}$$

where the offset voltage V_{offset,Σ_3} also considers measurement sensor offsets and any
other additional influences on the module voltage. Electrical resistances are usually
linear elements, but a current-dependent adaptation of the internal resistance and
offset, respectively, has been shown as advantageous. This can be caused by, e.g.,
inaccuracies in the battery cells, different ages of the installed battery cells, or the
battery cell used to identify the battery cell model. Nevertheless, the resulting model
can be denoted by

$$V_{module, \Sigma_4} = n_{cells} \cdot V_{cell} - R^{\pm}_{int, \Sigma_4} \cdot I_{module} + V^{\pm}_{offset, \Sigma_4} \qquad (2.29)$$

with

$$R^{\pm}_{int, \Sigma_4} = \begin{cases} R^{+}_{int, \Sigma_4} & \text{if } I_{module} > 0 \\ R^{-}_{int, \Sigma_4} & \text{if } I_{module} \leq 0 \end{cases}, \quad V^{\pm}_{offset, \Sigma_4} = \begin{cases} V^{+}_{offset, \Sigma_4} & \text{if } I_{module} > 0 \\ V^{-}_{offset, \Sigma_4} & \text{if } I_{module} \leq 0 \end{cases}.$$
$$(2.30)$$

Battery module measurements can then be used to identify the module parameters R_{int, Σ_i} and V_{offset, Σ_i}, respectively.

2.6   State of Charge Estimation

State estimation is required, if system states are not measurable. An observer is used to estimate the unmeasurable states based on the measurable states and a dynamic model of the process. In this context, the observability of the system must be given, which is the case if the observability matrix $\mathcal{O}$

$$\mathcal{O} = \left[\mathbf{C}^{\mathrm{T}} (\mathbf{CA})^{\mathrm{T}} (\mathbf{CA}^2)^{\mathrm{T}} \dots (\mathbf{CA}^{n_{states}-1})^{\mathrm{T}} \right]^{\mathrm{T}}, \qquad (2.31)$$

where n_{states} denotes the number of states in the state-space system with state matrix $\mathbf{A}$ and output matrix $\mathbf{C}$, has full rank.

The linear Kalman filter is an efficient recursive filter that is able to observe the internal states of a linear dynamic system by measurements corrupted with noise. It is of advantage that the filter is based on time-invariant models, which significantly reduces the computational complexity. In general, nonlinear processes are more challenging, since nonlinear estimation approaches need to be applied. A well-known nonlinear estimator is the extended Kalman filter, which is based on the local Jacobian of a nonlinear model and is therefore more complex than a linear Kalman filter. Note that the filter gain as well as the data-dependent local linearization of the EKF cannot be precalculated, which is disadvantageous for real-time application (Plett 2004b; Vasebi et al. 2007). The interacting multiple models' approach achieves a nonlinear estimation by running separate linear Kalman filters and switching between the filters based on a detection scheme using the validity probability of the underlying models (Mazor et al. 1998; Helm et al. 2012). Another similar nonlinear estimator, which is ideal in combination with LMN models, is the fuzzy observer (Chen et al. 1998; Senthil et al. 2006). Since each LLM of the LMN corresponds to an linear time-invariant dynamic system, the linear Kalman filter theory can be applied, and the global filter output can be obtained by weighted aggregation of the local filters (Simon 2003). Furthermore, stability of the fuzzy observer can be shown based on Lyapunov stability theory and is discussed in, e.g., Mayr et al. (2011b) and Tanaka et al. (1998). The fuzzy observer approach is therefore used in this book.

In the following, the general architecture of the SoC estimation scheme is presented before the development of the fuzzy SoC observer. Note that the SoC estimation approach is applicable for cell and module SoC estimation, respectively, and is therefore developed for the general case.

2.6.1 General Architecture of the SoC Observer Scheme

The estimation of the SoC of a battery can be obtained by two practical principles. First, measuring the open-circuit voltage leads to an accurate estimate of the SoC if the battery has been in standby mode for a sufficiently long period of time, but cannot be used during operation. The second possibility is to integrate the current, which can be denoted by

$$\text{SoC}(t) = \text{SoC}_{init} + \frac{1}{Q_{c,batt}} \int_0^t \eta_{batt,cou}(I_{batt}(v))\, I_{batt}(v)dv, \qquad (2.32)$$

where SoC_{init} is the initial SoC, $Q_{c,batt}$ is the battery capacity, and $\eta_{batt,cou}$ is the charge efficiency. Note that $\eta_{batt,cou}$ is usually 0.99 and can be neglected for the purpose of observer based SoC estimation. Nevertheless, the integration of the current leads to an acceptable initial SoC value that can be used in an advanced SoC observer scheme.

Following Hametner and Jakubek (2013), the estimation of the SoC can be done by automated nonlinear observer design using a fuzzy observer. In this context, the relative SoC estimation given by Eq. (2.32) should be included in the observer scheme. For that purpose, Eq. (2.32) is represented in the discrete formulation

$$\text{SoC}(k) = \text{SoC}(k-1) + \frac{t_{s,bms}}{Q_{c,batt}} I_{batt}(k), \qquad (2.33)$$

where $t_{s,bms}$ corresponds to the sampling time. An overview of the observer scheme is depicted in Fig. 2.11 exemplarily. As can be seen in the figure, the SoC is corrected by the observer, which obtains the simulated battery voltage by the LMN battery model discussed in Sect. 2.2. Any drift of the SoC due to extended operational time is compensated, since the SoC estimator is able to act during operation. Note that the estimation accuracy is mainly dependent on the battery model accuracy. In the next subsection, the SoC fuzzy observer is developed in detail.

2.6.2 SoC Fuzzy Observer Design

The automated nonlinear observer design requires the nonlinear LMN in state-space (SS) representation (Senthil et al. 2006), the state vector of which needs to be

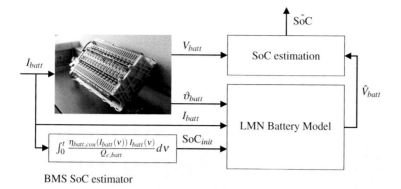

BMS SoC estimator

Fig. 2.11 SoC fuzzy observer scheme

augmented in order to consider the initial SoC from the current integration in the fuzzy observer. A correct transformation of the LLM (c.f. Eq. (2.2)) into the SS representation can be achieved by including the past outputs y and the previous state of charge $\mathrm{SoC}(k-1)$ in the augmented state vector $\mathbf{x}_{aug}$, which follows to

$$\mathbf{x}_{aug}(k) = \left[y(k-1) \; y(k-2) \; \ldots \; y(k-n) \; \mathrm{SoC}(k-1) \right]^{\mathrm{T}}. \qquad (2.34)$$

The parameter vector of the ith LLM can be denoted by

$$\boldsymbol{\vartheta}_i = \left[a_{1,i} \; \ldots \; a_{n,i} \; b_{11,i} \; \ldots \; b_{lm_l,i} \; \ldots \; b_{qm_q,i} \; b_{0,i} \right], \qquad (2.35)$$

where $a_{\cdot,i}$ and $b_{\cdot,i}$ denote the output and input parameters, respectively, while $b_{11,i} = b_{\mathrm{SoC},i}$ corresponds to the order-one SoC input parameter used in the LMN design (cf. Eq. (2.8)). Note that the LMN design described in Sect. 2.2.3 already includes the SoC as input variable, in order to be able to use the identified parameter for the augmented state variable in the design of the linear Kalman filters. Using these parameters, the state matrix $\mathbf{A}_i$ can be established by

$$\mathbf{A}_i = \begin{bmatrix} a_{1,i} & a_{2,i} & \ldots & a_{n,i} & b_{\mathrm{SoC},i} \\ 1 & 0 & \ldots & 0 & 0 \\ 0 & 1 & \ldots & 0 & 0 \\ \vdots & \vdots & \ddots & \vdots & \vdots \\ 0 & 0 & \ldots & 0 & 1 \end{bmatrix} \qquad (2.36)$$

and the input matrix $\mathbf{B}_i$ follows by

$$
\mathbf{B}_i = \begin{bmatrix} b_{21,i} & \cdots & b_{lm_l,i} & \cdots & b_{qm_q,i} & b_{0,i} \\ 0 & 0 & 0 & 0 & 0 & 0 \\ \vdots & \ddots & \vdots & \ddots & \vdots & \vdots \\ \frac{t_{s,bms}}{Q_{c,batt}} & 0 & 0 & 0 & 0 & 0 \end{bmatrix} ,
\tag{2.37}
$$

where $b_{0,i}$ corresponds to the local affine term (bias term). Note that the corresponding parameter of the SoC input $b_{11,i} = b_{\mathrm{SoC},i}$ is excluded from $\mathbf{B}_i$ since it is already considered in $\mathbf{A}_i$. The augmented input vector for the augmented observer SS model follows then to

$$
\mathbf{u}_{aug}(k) = \begin{bmatrix} u_1(k) & u_1(k-1) & \cdots & u_q(k-m_q) & 1 \end{bmatrix}^{\mathrm{T}} ,
\tag{2.38}
$$

where u_i corresponds to the actual and past elements of the model inputs. At this point it is important to mention that the integrator needs to be adapted if a normalization of the LMN is used. In case of scaled parameters, the integrator increment can be transformed by a corresponding additional term in the last row and column of the matrix $\mathbf{B}_i$.

Using Eqs. (2.34)–(2.38), the augmented state equation

$$
\mathbf{x}_{aug}(k) = \sum_{i=1}^{M} \Phi_i(k-1) \left\{ \mathbf{A}_i \mathbf{x}_{aug}(k-1) + \mathbf{B}_i \mathbf{u}_{aug}(k) \right\}
\tag{2.39}
$$

leads to the global model output

$$
y(k) = \mathbf{C} \mathbf{x}_{aug}(k)
\tag{2.40}
$$

with

$$
\mathbf{C} = \mathbf{C}_i = \begin{bmatrix} 1 & 0 & \ldots & 0 & 0 \end{bmatrix} .
\tag{2.41}
$$

Based on these equations, the local steady-state Kalman filters can be designed for each local linear model, while the global filter is obtained by weighted aggregation of the individual local estimates. The state estimate $\hat{\mathbf{x}}_{aug}$ of the global filter can be determined based on the gain matrix $\mathbf{K}_i$ by

$$
\hat{\mathbf{x}}_{aug}(k) = \sum_{i=1}^{M} \Phi_i(k-1) \left\{ \mathbf{x}^*_{aug,i}(k) + \mathbf{K}_i \left[y(k) - \hat{y}(k) \right] \right\} ,
\tag{2.42}
$$

with

$$
\mathbf{x}^*_{aug,i}(k) = \mathbf{A}_i \hat{\mathbf{x}}_{aug}(k-1) + \mathbf{B}_i \mathbf{u}_{aug}(k) ,
\tag{2.43}
$$

where

$$\hat{y}(k) = \sum_{i=1}^{M} \Phi_i(k-1) \mathbf{C} \mathbf{x}_{aug,i}^{*}(k). \tag{2.44}$$

The gain matrix $\mathbf{K}_i$ is calculated by

$$\mathbf{K}_i = \mathbf{A}_i \mathbf{P}_i^T \mathbf{C}^T \left(\mathbf{C} \mathbf{P}_i^T \mathbf{C}^T + \mathbf{R}^T \right)^{-1}, \tag{2.45}$$

where $\mathbf{P}_i$ refers to the solution of the discrete-time algebraic Riccati equation (DARE)

$$\mathbf{A}_i \mathbf{P}_i \mathbf{A}_i^T - \mathbf{P}_i - \mathbf{A}_i \mathbf{P}_i \mathbf{C}^T \left(\mathbf{C} \mathbf{P}_i \mathbf{C}^T + \mathbf{R} \right)^{-1} \mathbf{C} \mathbf{P}_i \mathbf{A}_i^T + \mathbf{Q} = 0. \tag{2.46}$$

In Eq. (2.46), the matrices $\mathbf{R}$ and $\mathbf{Q}$ reflect the covariance matrices of the measurement and process noise, respectively. The Kalman filter theory assumes $\mathbf{R}, \mathbf{Q}$ to be known, but often the covariance matrices are unknown and $\mathbf{R}$ and $\mathbf{Q}$ are used as tuning parameters (Hametner and Jakubek 2013). For the application of SoC estimation in batteries, the terminal voltage measurement noise $\mathbf{R}$ can be found by experiments. The process noise $\mathbf{Q}$, similar to Vasebi et al. (2007), is used to tune the filter and no correlation between the different elements in $\mathbf{Q}$ is assumed.

Chapter 3
Results for BMS in Non-Road Vehicles

Abstract The battery management system of batteries used in non-road vehicles needs to be able to provide an accurate value of the SoC, even during operation if high dynamic currents act on the battery. This can be achieved based on the generically applicable methodologies proposed in Chap. 2. In this chapter, the proposed concepts and methodologies are validated using measurements from real battery cells with different cell chemistries as well as a battery module. The battery module is established with battery cells, which are separately available to be investigated with cell measurements.

Keywords Lithium ion batteries · Nonlinear system identification · Optimal model based design of experiments (DoE) · State observer

In the following, first, the measurement procedures and hardware are specified. Second, the results for the nonlinear battery cell model identification are discussed. Third, the results for the temperature and battery module model are validated and at the end, the estimation accuracy achieved with the SoC estimator introduced in Sect. 2.6 is demonstrated.

3.1 Generation of Reproducible High Dynamic Data Sets

The reliability of precise battery models depends on the data set used to identify the model parameters. Due to this reason, the reproducibility of measurements is essential for the whole methodology, since only reproducible measurements can verify the model accuracy. In the following, the measurement procedure is discussed and the measurement hardware is described. At the time, when battery cell measurements were required, an affordable and sufficiently high dynamic battery cell tester was not available, and therefore a cell tester hardware has been developed. Measurements on a real battery module could be achieved using an adapted battery simulator available at the test bed also used to measure the hybrid powertrain (cf. Chap. 5). In the following, the measurement procedures are introduced before the developed battery cell tester and the module tester are described in detail.

© The Author(s) 2016
J. Unger et al., *Energy Efficient Non-Road Hybrid Electric Vehicles*,
SpringerBriefs in Applied Sciences and Technology,
DOI 10.1007/978-3-319-29796-5_3

3.1.1 Measurement Procedures

Electrochemical batteries show that the previous excitation has influence on the voltage behavior of the cells. Reproducibility can only be guaranteed if the initial battery conditions are uniquely defined in advance to the measurements. Unexpected and undefined appearing effects are avoided and all short-term history of the battery cells (cell conditioning) is erased. It has been shown that the following procedure enables to compare measurements as well as different battery cells and chemistries, respectively, with each other (Unger et al. 2014):

1. Initial capacity check at 25 °C
2. Set temperature of climate chamber
3. Fully charge the battery cell
4. Discharge until initial SoC is reached
5. Apply excitation signal
6. Repeat 3 to 5 until all excitation signals are recorded
7. Repeat 2 to 6 until all temperatures are recorded

Since the exact capacity of a battery cell is unknown in advance, the initial capacity check (see enumeration 1) is applied only once before any dynamic measurements are taken. Eight charge/discharge cycles with different C-rates from $4C$ to $1C$ are applied to the battery cell, due to which the short-term history is reduced and the true capacity can be calculated. Due to the high C-rates, the full capacity of a battery cell cannot be exploited, because of which only the last cycle is used to calculate the capacity. The quotient of discharge and charge capacity can additionally be used to obtain a charge efficiency for each cycle. Any battery cell measurements obtained in this book are based on this procedure.

In case of battery modules, the capacity check is more difficult due to the cell balancing. Nevertheless, in order to achieve the best reproducibility possible, the battery module is fully charged and balanced before the battery module is discharged to the demanded initial SoC. Prior to the measurements, a conditioning cooling circuit established the demand temperature of the battery module.

3.1.2 Test Hardware for Battery Cells

High dynamic current excitation signals such as real load profiles are not state of the art yet, which implicates that reasonably priced high dynamic battery cell testers are unavailable. Due to this reason, a battery cell tester is developed based on a Höcherl&Hackl source/drain module for closed-loop current control in real time and a National Instruments USB data acquisition board for the measurements. A LabView control software executed the measurement procedure, kept any physical constraints, and monitored battery states and safety issues. The voltage range is -1– $10\,V$, and the current range is $\pm240\,A$. A Vötsch climate chamber, also triggered

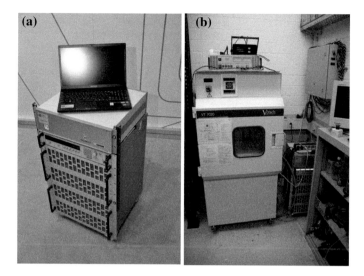

Fig. 3.1 Battery cell tester (*left*) and climate chamber (*right*) used to test different cell chemistries and temperatures

by the LabView control software, provided an ambient temperature between −20 and 60 °C (relevant temperature range for non-road application is between 12.5 and 35 °C). Note that the system is designed for up to 10 kHz with step response time constants of the source/drain module for less than 200 μs. Due to the slow system behavior of battery cells, the measurement duration is very long, and an enormous amount of data is generated if the sampling time is chosen too high. Any voltage response from the applied excitation signals and the cell temperature are therefore recorded with a sampling rate of 100 Hz. In Fig. 3.1, the battery cell tester (left) and the climate chamber (right) are depicted.

3.1.3   Test Hardware for Battery Modules

A battery simulator (KS BattSim) from Kristl, Seibt & Co GmbH, usually designated to emulate battery models in real time, is adapted to test the available real battery module. The battery simulator has a voltage range from 500 to 750 V and a current range of ±250 A. These specifications are sufficient for the battery module measurements and the emulation of the battery module during the hybrid powertrain test bed measurements. The current excitation signal is directly programmed into the control unit of the battery simulator, and an implementation of high dynamic excitation signals is possible. Depending on the purpose of the battery simulator, the control interface is changed.

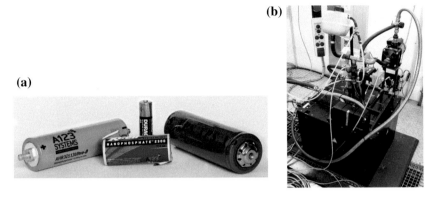

Fig. 3.2 Exemplary battery cells commonly used in HEV compared to a commercial AA-type battery cell (*left*) and real battery module (*right*)

3.2 Battery Cells and Battery Module Specifications

The presented methodologies are data-based approaches that are generically applicable to any battery chemistry. In order to show this benefit, battery cells with different cell chemistries are investigated in this book. The three battery cells and one battery module under investigation are introduced in the following.

Cell A is a prismatic type, energy cell based on lithium-polymer chemistry, which has a nominal voltage of 3.7 V. Cell B and C are cylindrical type, power cells based on lithium-iron-phosphate cell chemistry and have a nominal voltage of 3.3 V. The capacity of Cell A is 40 A h, while Cell B and Cell C have a capacity of 4.4 A h and 1.1 A h, respectively. In Fig. 3.2 (left), similar battery cells commonly used in HEV are depicted exemplarily and compared with a standard AA-sized primary battery cell. The battery module is built by 192 series and 2 parallel connected cells (192S2P-configuration) of type Cell B, and has a nominal voltage of 630 V and a maximal current of ± 200 A (see Fig. 3.2 (right)). Passive cell balancing is established between the battery cells, but balancing does not start until a 15 minute standby time of the battery. All battery cells and the battery module allow a temperature range for charge and discharge, respectively, between 0 and 40 °C, although the investigated and relevant temperature range in non-road application lies between 12.5 and 35 °C.

3.3 Training Data for Battery Cell Models

The methodology of optimal model-based DoE, described in Sect. 2.3, provides excitation signals, which minimize the parameter variances of the identified battery models. In this book, these signals are used as training data and should be presented exemplarily for battery cell B in the following. Figure 3.3 shows the result for the

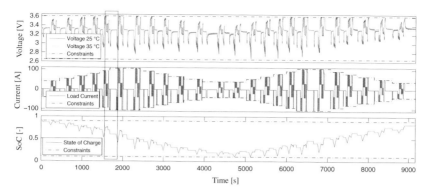

Fig. 3.3 Resulting optimal battery excitation for Cell **B**, measured at two different temperatures. See Fig. 3.4 for a detailed view

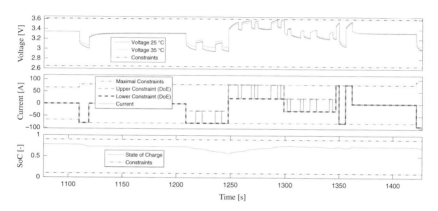

Fig. 3.4 Detailed enlargement of the marked region in Fig. 3.3 including the current constraints applied in the optimization

obtained optimal excitation signal with the recorded voltage response of battery cell **B**. The current signal and current constraints are depicted in Subplot two, where currents with more than 100 A can be observed. Cell **B** has a capacity of 4.4 A h, which results in a C-rate of more than $22.5C$. Nevertheless, the corresponding voltage responses for the two depicted temperatures (see Subplot one) and the SoC (see Subplot three) keep the constraints. Note that even though the constraints of voltage and SoC are indirectly considered, any constraints are met, and the entire relevant range of the SoC is covered.

In Fig. 3.4, one extended excitation sequence is shown in more detail. Subplot two shows that the optimization considered the current constraints and the maximal current depending on the SoC, respectively. Furthermore, it can be seen that high dynamic current is obtained between the lower and upper current constraints (green and red lines), which is the optimized part of the optimal model-based DoE methodology. However, the voltage response depicted in Subplot one shows that the degree of freedom has not been fully exploited due to the linear reference model. Note that the spread of the current over the dynamic range may also include more intermediate steps, if a nonlinear reference model is used (Unger et al. 2012a).

3.4 Validation of Battery Cell Model Accuracy

The results obtained with LMN battery models are presented in this section. First, the improvements due to optimal excitation signals are discussed based on two different LMN battery models. Second, the generic applicability to different cell chemistries and the influence of the mentioned nonlinearities are demonstrated. Finally, the best LMN battery model is further discussed in terms of dynamic accuracy.

3.4.1 Battery Model Quality Improvement with Optimal DoE

The accuracy of data-based LMN is significantly dependent on the used training data. For this reason, two local model networks (models Γ_1 and Γ_2) with the same configuration are identified by conventional DoE and optimal model-based DoE, respectively, to show this dependency. The benefit can better be seen on a simplified LMN structure, which is chosen to $\mathcal{Z}_{12} = \begin{bmatrix} z_{\text{SoC}} & z_{\text{Hyst}} \end{bmatrix}$ as partition space and $\mathcal{Q}_{12} = [u_{\text{Current}}]$ as input space of the models. Since the ambient temperature is kept constant at 22.5 °C, z_{Temp} is excluded in $\mathcal{Z}_{12}$, and z_{SoC} and u_{Relax} are neglected in $\mathcal{Q}_{12}$ in order to exclusively show the effect of the training data. Following the methodology by Jackey et al. (2013), a good compromise for the model order is found by analyzing the mean squared error of a linear model for different selections of the order and choosing a suitable model order between complexity and accuracy. The model orders of current and voltage are finally set by $m_{\text{Current}} = n_{\text{Voltage}} = 5$.

An LMN model with 10 local linear models is used to find a suitable choice for the kernel function sharpness of the SoC $k_{\sigma,\text{SoC}}$. The given configuration of the partition space leads to a good compromise by $k_{\sigma,\text{SoC},12} = 0.75$, which is obtained by comparing different selections of $k_{\sigma,\text{SoC}}$ with each other. Note that smoothness and strict partitioning is controlled by $k_{\sigma,\text{SoC}}$, since the overlapping of the validity functions is increased with a larger $k_{\sigma,\text{SoC}}$ value. A smooth separation between discharge and charge behavior of z_{Hyst} is not physically reasonable, due to which a desired sharp separation is obtained by $k_{\sigma,\text{Hyst}} = 0.05$. The threshold for the number of LLM is predefined by the critical real-time limit of the battery emulator control unit (Beckhoff Industrial PC C6515), which is capable of $M = 30$ as is obtained by test runs.[1] Table 3.1 shows the summary of the LMN configuration parameters for models Γ_1 and Γ_2.

---

[1]8–12 % CPU usage of battery emulator.

Table 3.1 Collection of the parameter configuration used in battery models Γ_1 and Γ_2

Structure of LMN	Values of parameter configuration
$\mathcal{Q}_{12} = [u_{\text{Current}}]$ $\mathcal{Z}_{12} = \begin{bmatrix} z_{\text{SoC}} & z_{\text{Hyst}} \end{bmatrix}$	$m_{\text{Current}} = n_{\text{Voltage}} = 5$ $k_{\sigma,\text{SoC},12} = 0.75$ $k_{\sigma,\text{Hyst}} = 0.05$ $M = 30$

In Fig. 3.5, the raw training data for model Γ_1 is depicted, where conventional DoE for Cell $\boldsymbol{C}$ is used to obtain the excitation signal. Hu et al. (2009b) proposed a step profile training data, which is similar to the one used. Note that the conventional DoE is more dynamic than other excitation signals in the literature (Unger et al. 2012a) and is therefore used for comparison with the proposed experiment design. However, the step profile is established by alternating intermediate current steps in charge and discharge directions, and the variation of the step durations realizes to cover the entire SoC range. Disadvantageous is that the DoE is very strict in terms of voltage, since voltage behavior is not considered. Expert knowledge needs to be used to limit the current in order to avoid voltage violations. Nevertheless, the applied current constraints can be seen in the second subplot, which ensure that current, voltage, and SoC constraints are met, although full current capabilities of the battery cell are not exploited by the step profile. Subplot one includes the training result obtained for the battery model Γ_1.

Model Γ_2 is identified using an optimal excitation signal obtained with the methodology proposed in Sect. 2.3 and adapted for Cell $\boldsymbol{C}$. The optimal model-based DoE is depicted in Fig. 3.6, where the training result of battery model Γ_2 is included in the first subplot. Subplot two depicts the optimal excitation signal including the applied constraints, which indirectly consider the voltage and SoC limits, respectively. In comparison to the conventional DoE in Fig. 3.5, the optimal excitation signal has superior dynamics and higher currents, while all constraints are met.

In order to validate the obtained battery models Γ_1 and Γ_2 in terms of dynamics and high currents occurring in non-road applications, a repeated real (current-) load cycle

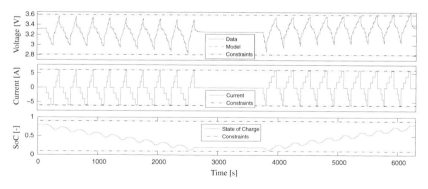

Fig. 3.5 Measurements from Cell $\boldsymbol{C}$ at 22.5 °C used to training LMN model Γ_1

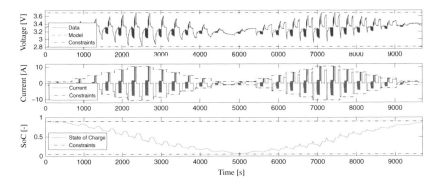

Fig. 3.6 Resulting optimal excitation of Cell C at a temperature of 22.5 °C to training LMN model Γ_2

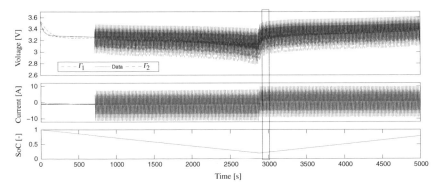

Fig. 3.7 Accuracy of models Γ_1 and Γ_2 shown on the simulation of the SoC validation signal for Cell C at 22.5 °C. See Fig. 3.8 for a detailed view

is used, which does not change the SoC on average. Alternatingly raising/lowering the mean current value of the cycle forces the SoC to pass the entire SoC range, due to which the signal is referred to as *SoC validation signal*. Note that in the following, the maximal current capabilities of the battery cells under investigation are exhausted by scaling the cycle to the maximal current allowed for the specific cells. Figure 3.7 depicts the obtained simulation results for models Γ_1 and Γ_2. In Subplot one, the voltage response of the battery and the simulated voltages of the models are shown. Subplot two and three show the current and the SoC trajectory, respectively. The ambient temperature during the experiment is kept constant at 22.5 °C.

Both models consider the nonlinear relationship between SoC and battery cell voltage, which is observable by comparing the envelope curves of the maximal voltage values in Subplot one of Fig. 3.7. Nevertheless, a significant average error of model Γ_1 (56.25 mV) can be seen compared to the error of model Γ_2 (5.44 mV). The marked region in Fig. 3.7 is depicted in detail in Fig. 3.8.

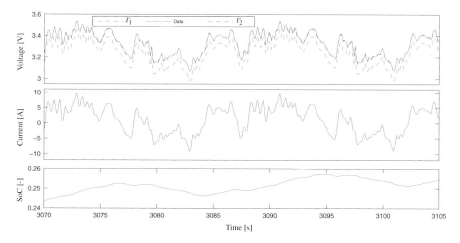

Fig. 3.8 Model accuracy of models Γ_1 and Γ_2 shown on a detailed view of the real load cycle simulation

In Fig. 3.8, one cycle is enlarged to show the dynamic behavior of models Γ_1 and Γ_2. Behaviors are generally similar, but model Γ_1 shows a large offset compared to Γ_2, which is observable in the first subplot. One reason is the lack of high dynamic excitation in the step excitation signal used to train model Γ_1. Due to the optimal model-based DoE, an optimal excitation signal is obtained, which is able to significantly increase the model accuracy for high dynamic excited battery cells. Even though C-rates above $9C$ occur in the validation signal, satisfactory model accuracy is obtained. For this reason, exclusively optimal excitation signals are used in the following to train the battery models.

3.4.2 Comparison of Battery Cell Models with Different LMN Structures and Cell Chemistries

The influence of optimal excitation signals on the LMN battery model quality is presented in the previous section. In the following, the influence of hysteresis, relaxation, and temperature input on the LMN structure as well as the applicability of the LMN approach to different battery cell chemistries is discussed in detail. To this end, three different LMN battery model structures (Γ_3, Γ_4, Γ_5) are validated using two different validation signals (*temperature validation signal* and *SoC validation signal*) and two different cell chemistries (Cell **A** with lithium-polymer and Cell **B** with lithium-iron-phosphate chemistry). The achieved results are then compared with each other and interpreted.

Training data for the three models is obtained by creating optimal excitation signals for each cell chemistry separately and measuring the voltage response at 10 different temperatures between 12.5 and 35 °C in 2.5 °C steps. Figure 3.3 shows the

Table 3.2 Collection of the parameter configuration used in battery models Γ_3, Γ_4 and Γ_5

Structure of Γ_3, Γ_4 and Γ_5 LMNs	Configuration Parameters for Γ_3, Γ_4 and Γ_5
Γ_3: $Q_3 = [z_{SoC}\ u_{Current}]$, $Z_3 = \begin{bmatrix} z_{SoC} & z_{Current} & z_{Temp} \end{bmatrix}$ Γ_4: $Q_4 = [z_{SoC}\ u_{Current}]$, $Z_4 = \begin{bmatrix} z_{SoC} & z_{Hyst} & z_{Temp} \end{bmatrix}$, Γ_5: $Q_5 = [z_{SoC}\ u_{Current}\ u_{Relax}]$, $Z_5 = \begin{bmatrix} z_{SoC} & z_{Hyst} & z_{Temp} \end{bmatrix}$	$m_{SoC} = 1$, $k_{\sigma,SoC,345} = k_{\sigma,Temp} = 0.6$, $m_{Relax} = 3$, $k_{\sigma,Hyst} = k_{\sigma,Current} = 0.05$, $m_{Current} = n_{Voltage} = 5$, $M = 30$

optimal excitation signal of Cell **B** at two different temperatures exemplarily for all cell chemistries, which are then merged to one training data set (raw identification data) for Cell **A** and **B**, similarly.

In Table 3.2, the LMN structures and configuration parameters of the models explained in the following are summarized. The purpose of the simplest model (Γ_3) is to provide a reference to show the influence of hysteresis and relaxation input to the model structure, which are not considered in Γ_3. However, the temperature is considered in the partition space, which follows by $Z_3 = \begin{bmatrix} z_{SoC} & z_{Current} & z_{Temp} \end{bmatrix}$ and the input space, including the SoC as mentioned in Sect. 2.2.3, is defined by $Q_3 = [z_{SoC}\ u_{Current}]$. Note that due to the excluded hysteresis input, the corresponding polarization of the battery cell is not indicated through the current $z_{Current}$ at standby, since $z_{Current}$ is only interpolated between charge and discharge behavior. The current is therefore initially partitioned between positive and negative current instead of charge and discharge mode, but nevertheless the current effects are considered in model Γ_3.

Model Γ_4 is improved compared to model Γ_3 by replacing the current input with the hysteresis input, which leads to a partition space $Z_4 = \begin{bmatrix} z_{SoC} & z_{Hyst} & z_{Temp} \end{bmatrix}$ and an input space $Q_4 = Q_3$. At this point, it is important to mention that the complexity of the model is unchanged in this case. An increase in model complexity is obtained in model Γ_5, which includes the relaxation input in the input space. Hence, the input space can be denoted by $Q_5 = [z_{SoC}\ u_{Current}\ u_{Relax}]$, and the partition space remains the same ($Z_4 = Z_5$). Note here that any nonlinear and/or electrochemical effects mentioned in this book are considered in model Γ_5.

In terms of the configuration parameters, the objectives remain the same as for the models Γ_1 and Γ_2 and can therefore directly taken from Table 3.1. Therefore, the parameters for models Γ_3, Γ_4, and Γ_5 follow to $m_{Current} = n_{Voltage} = 5$, $k_{\sigma,Hyst} = 0.05$, and $M = 30$, while $k_{\sigma,Current} = k_{\sigma,Hyst} = 0.05$ is set for the sharpness factor of the current input to achieve a sharp separation between charge and discharge behaviors. In case of the optimal kernel function sharpness of the SoC and temperature, $k_{\sigma,SoC,345} = k_{\sigma,Temp} = 0.6$ slightly increases the smoothness of the validity function and is therefore chosen for the temperature input. The SoC input has no dynamic influence on the system behavior, for which reason the order of the input is set straight forward to $m_{SoC} = 1$. Because $filt(\cdot)$ of the relaxation input is established using a third-order low-pass filter, the order within the LMN model is also set straight forward to $m_{Relax} = 3$.

The validation of the battery models is achieved using the two mentioned validation signals for temperature and SoC, while the *SoC validation signal* is already introduced beforehand and slightly improved. Aim of the temperature validation is to show the model accuracy in case of temperature changes and different SoC values. Due to this reason, the ambient temperature is heated up to the upper level of 32 °C and cooled back to the lower level of 18 °C, while the aforementioned real load cycle is continuously applied to the battery cell (Unger et al. 2014). This procedure is repeated for different SoC levels and is referred to as *temperature validation signal*.

The *SoC validation signal*, as introduced earlier in this section, is furthermore improved to strengthen the significance of the validation by alternatingly raising/lowering the current mean value in between the repetitions of the cycle. Due to the more often changed average charge direction, the dynamic behavior is more excited and therefore proves the battery model with more significance, though the entire SoC range is covered. In the next subsection, the *temperature validation signal* and *SoC validation signal* are described in terms of the dynamic behavior and depicted in Figs. 3.9 and 3.11, respectively.

Based on the mean squared error and a normalized root mean squared error (NRMSE), the model accuracy can be evaluated. The NRMSE can be denoted by

$$e_{NRMSE,\%} = \sqrt{\frac{1}{N_s} \sum_{i=1}^{N_s} \left(\frac{y_i - \hat{y}_i}{\max(y) - \min(y)} \right)^2} \cdot 100\,\%, \tag{3.1}$$

where y is the measured output, $\hat{y}$ is the simulated output, and N_s is the number of samples. Note that the different cell chemistries have different nominal voltages, due to which a normalization of the measured output values in Eq. (3.1) is required in order to allow a direct comparison between the different battery model accuracies. Table 3.3 presents the MSE and NRMS values for the different LMN architectures and cell chemistries.

In Table 3.3, a continuous decrease in error values is observable from model Γ_3 to model Γ_5. Reasons are the included hysteresis input (Γ_4) and the increased complexity (Γ_5) similar for both cell chemistries. The interpolation between the charge/discharge behavior at standby current in model Γ_3 is not physically appropriate, while in contrast, the corresponding polarization is indicated through the hysteresis input in model Γ_4 and Γ_5. In terms of the electrochemical behavior of battery cells, models Γ_4 and Γ_5 are more physically appropriate, and therefore the model accuracies are improved.

The time constants of electrochemical batteries are usually very different, due to which a high sampling rate is required for the measurements, since otherwise the fast dynamic behavior is not considered in the recorded data. A fast sampling rate causes a highly correlated data set, though, and leads to numerical problems due to ill-conditioning at the parameter estimation (Billings and Aguirre 1995; Ljung 1998). This phenomenon is referred to as redundance (Billings and Aguirre 1995). The used sample time of 100 Hz makes it very difficult to precisely identify the corresponding relaxation time constant within the dynamic behavior. Therefore, the relaxation input

Table 3.3 Accuracy comparison of models Γ_3, Γ_4 and Γ_5 based on error values obtained for Cell **A** and Cell **B**

		Temp. Val. Sig.	SoC Val. Sig. 20 °C	SoC Val. Sig. 25 °C	SoC Val. Sig. 30 °C
MSE Cell **A**	Γ_3	2.935e^{-4}	3.711e^{-4}	–	–
	Γ_4	2.668e^{-4}	3.559e^{-4}	–	–
	Γ_5	2.447e^{-4}	1.431e^{-4}	–	–
MSE Cell **B**	Γ_3	7.592e^{-4}	5.652e^{-4}	4.622e^{-4}	3.633e^{-4}
	Γ_4	6.585e^{-4}	4.811e^{-4}	4.094e^{-4}	3.393e^{-4}
	Γ_5	5.236e^{-4}	3.386e^{-4}	3.109e^{-4}	2.496e^{-4}
NRMSE Cell **A**	Γ_3	2.059 %	2.537 %	–	–
	Γ_4	1.963 %	2.485 %	–	–
	Γ_5	1.752 %	1.438 %	–	–
NRMSE Cell **B**	Γ_3	2.671 %	2.933 %	2.928 %	2.847 %
	Γ_4	2.487 %	2.706 %	2.756 %	2.752 %
	Γ_5	2.218 %	2.270 %	2.402 %	2.360 %

of model Γ_5 is used to provide the required information of the slow relaxation time constant. As is observable in the results, this improves the accuracy significantly.

Based on Table 3.3 and the corresponding MSE/NRMSE values, the best model accuracy of the three discussed models (Γ_3, Γ_4, Γ_5) is achieved by model Γ_5, which is discussed in detail in the following.

3.4.3 Dynamic Accuracy of the LMN Battery Models

Previously, only the error values were used to evaluate the battery model accuracy. In the following, the dynamic accuracy achieved with the introduced battery model Γ_5, which considers all relevant electrochemical effects, is discussed in detail in terms of relaxation, hysteresis, and temperature influence. The dynamic behavior of the measured cell chemistries is all similar, due to which the resulting plots are depicted for Cell **B** only. Figure 3.9 shows the temperature validation signal, where SoC and temperature trajectory are depicted in Subplot three and four, respectively.

The measured voltage response and the simulated cell voltage using model Γ_5 can be seen in Subplot one in Fig. 3.9. The dependence of the model output on the ambient temperature is clearly observable. A changing time constant of the relaxation effect, which is not considered in the constant time constant of the filter input, causes a model mismatch at the beginning of the dynamic excitation after discharging/charging to the different SoC levels. Nevertheless, following Table 3.3, the model accuracy is increased significantly due to the relaxation input. Since high battery currents (more than $20C$) occur in non-road applications, the training data is focused to high dynamically excite the battery behavior, which leads to a lack of sufficient information at low constant current (below $2C$). This is observable at the small model mismatch

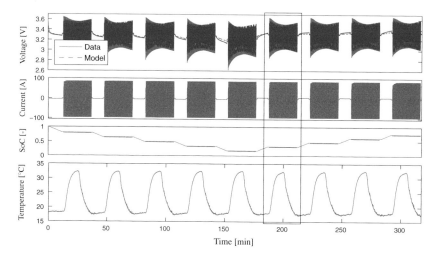

Fig. 3.9 Accuracy of model Γ_5 shown on the simulation of the temperature validation signal for Cell **B**. See Fig. 3.10. Detail (a) for an enlarged view

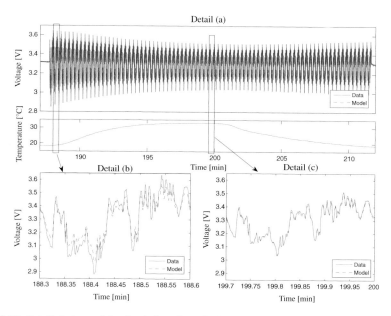

Fig. 3.10 Detailed views of the simulation of one SoC level during a temperature change between 18 and 32 °C

during transition to the different SoC levels using low constant currents. Note that the continuous change between charge and discharge prevents from observing the hysteresis effect directly in Fig. 3.9. Influence and benefit due to the hysteresis input are verified explicitly by the error values in Table 3.3, though. Figure 3.10 shows an enlarged view of the marked region (Detail (a)) in Fig. 3.9.

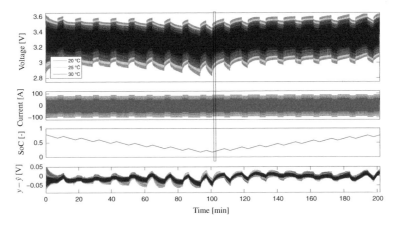

Fig. 3.11 SoC validation signal simulation results for Cell **B** at three temperatures (20, 25, and 30 °C). See Fig. 3.12 for an detailed view

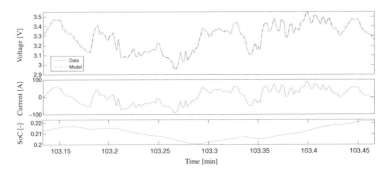

Fig. 3.12 Detailed view of the accuracy of model Γ_5 at 30 °C during one load cycle at Cell **B**

In Fig. 3.10, one load cycle is shown in an enlarged view. Detail (a) depicts the voltage and temperature signal, while Detail (b) and Detail (c) show the corresponding enlargement of the marked regions in Detail (a). The model dynamics, as can be seen, are sufficient for non-road applications. A comparison between Detail (b) and (c) shows the slightly bigger voltage error in Detail (b) due to the mentioned changing time constant of the relaxation effect.

Figure 3.11 presents the resulting SoC validation signal at a constant ambient temperature of 20, 25, and 30 °C, where the temperature influence on the voltage behavior is clearly observable.

Subplot four in Fig. 3.11 shows the invariant model error for the different temperatures, which never exceed a maximum error of 90 mV and stay within a small tolerance tube. Hence, it can be concluded that the SoC and temperature are considered in the battery model, since the error does not depend on the changing SoC or temperature. In Fig. 3.12, the marked region in Fig. 3.11 is depicted in detail, where one real load cycle with current rates above $20C$ at 30 °C ambient temperature can be seen.

The nonlinear behavior at low SoC is especially distinctive. In Fig. 3.12, it can be seen that although the SoC is low and high current values (above $20C$) occur, the LMN battery model accurately represents the underlying strong nonlinear behavior of the battery cell.

3.5   Battery Cell Temperature Model Accuracy

In order to simulate the battery model without measured signals, the temperature is required as model input. The simple temperature model proposed in Sect. 2.4 is able to provide sufficient accuracy for the simulation of the cell temperature. Based on the same training data used for the battery cell identification, the cell thermal capacity and heat transfer coefficient can be identified. For Cell B, the temperature model parameters are obtained to $c_p = 833.34 \frac{Ws}{K}$ and $h_{out} = 3.686 \frac{W}{K}$, respectively, which leads to the temperature model accuracy shown in Fig. 3.13. The first row of subplots shows the measured and simulated cell temperature, respectively, for the optimal excitation signal, the SoC validation signal, and the temperature validation signal. In the second row of subplots, the corresponding errors are depicted, while in the third row, the ambient temperature is shown. As can be seen is that the model quality is accurate, while only the transient temperature changes of the temperature validation signal show higher errors. This is caused by the external mounted temperature sensor, which is more influenced by the strong gradient of the ambient temperature in case of the temperature validation signal. Nevertheless, these ambient temperature gradients usually do not appear in real non-road vehicles, due to which the temperature model is legitimized. Note that the two error peaks in the SoC validation signal result from the merging of the validation signals and are observable at the non-realistic steps in the ambient temperature signal.

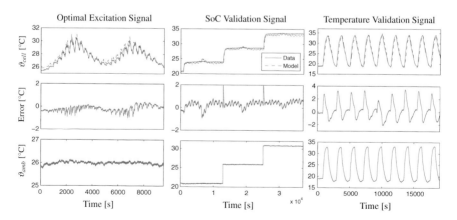

Fig. 3.13 Result obtained for the temperature model of Cell B

3.6 Battery Module Model Accuracy

The used battery module is built with $n_{cells} = 192$ series and $p_{cells} = 2$ parallel connected cells of type Cell $\boldsymbol{B}$. Hence, the obtained LMN battery cell model Γ_5 (cf. Table 3.2) can be used for the simulation of the battery module behavior, which is discussed in the following. Note that the balancing is activated after 15 minutes in standby mode and thus has no influence in the following.

Due to limited time capacities at the test bed, only step profiles are measured at three different current levels to create a set of training data for the battery module parameter identification. Nevertheless, the training results showed that the step profile with the full current coverage is sufficient. In Fig. 3.14, the used step profile is depicted.

In the first subplot of Fig. 3.14, voltage behavior as well as the training results is depicted for the four proposed module models. The corresponding current and SoC signals are shown in the second and third subplots, respectively. In the last subplot, the absolute value of the voltage difference between measurement and model (error) is shown. It can be observed that the error values are very small for all model approaches. An enlarged view of the training data is depicted in Fig. 3.15.

The error of the battery models is again depicted in the last subplot of Fig. 3.15. The significantly higher error of the simple model Σ_1 is clearly observable; it does not consider the internal resistance of the battery model. The other models are similar to each other, which indicates that the internal resistance is the major influencing effect beside the electrochemical effects considered by the LMN battery cell model. In order to validate the battery module models, a SoC validation signal is generated

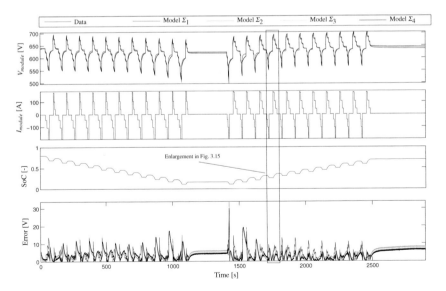

Fig. 3.14 Training data for battery module models

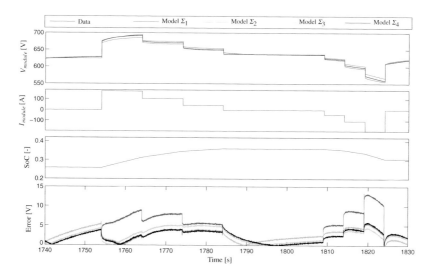

Fig. 3.15 Enlarged view of the marked region in Fig. 3.14

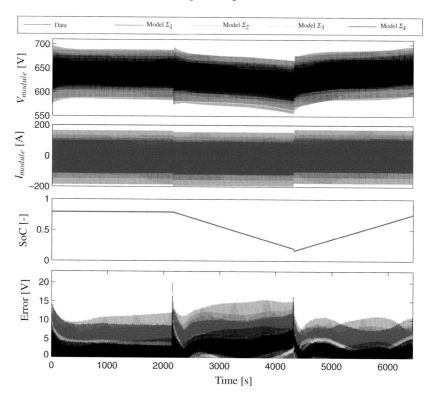

Fig. 3.16 Validation data for battery module models

that is based on the scaled real load cycle as introduced for the battery cell validation. The repeated real load cycles are superposed by zero, negative, and positive constant currents, respectively, to obtain the system behavior across the entire SoC range. Figure 3.16 depicts the SoC validation signal used for the battery module.

Subplot one of Fig. 3.16 shows the measured as well as simulated voltages. The second subplot shows the current signal, where one can see that the full current range is exploited. In Subplot three, the SoC profile is shown, which passes the entire range of SoC used in non-road vehicles. The last subplot shows the model errors, where the difference between model Σ_1 and the other models Σ_2, Σ_3, and Σ_4 is clearly observable. Note that the reason for the peak errors at the transitions results from data set merging, since a continuous measurement of the entire SoC validation signal is not supported by the battery tester. Nevertheless, Fig. 3.17 shows an enlarged view of the validation signal.

As can be seen in Fig. 3.17, all models are able to depict the high dynamic behavior of the battery module, while only model Σ_1 has a slightly larger mismatch. Although a current of almost ± 200 A is applied to the battery module, the accuracy is superior. Nevertheless, model Σ_4 has a slightly better overall accuracy compared to the other models. This is also observable in the MSE and NRMSE (cf. Eq. (3.1)) values obtained for the SoC validation signal as shown in Table 3.4.

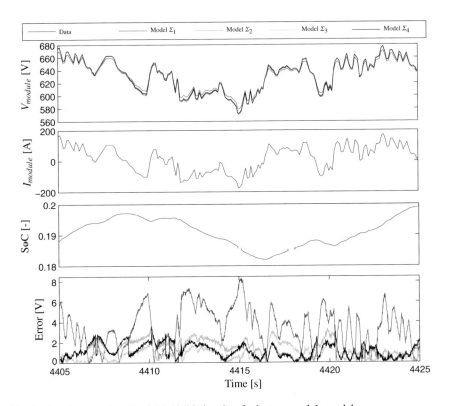

Fig. 3.17 Enlargement of Fig. 3.16: Validation data for battery module models

Table 3.4 Battery module SoC validation signal error values for the LMN battery module models Σ_1, Σ_2, Σ_3 and Σ_4

Criterion	MSE	NRMSE (%)
Model Σ_1	27.8851	2.5962
Model Σ_2	7.5727	1.3529
Model Σ_3	6.9949	1.3003
Model Σ_4	6.5871	1.2618

Table 3.4 shows the slight increase in accuracy due to the considered voltage off-set and current-dependent parameters, respectively. Nevertheless, a NRMSE clearly below 2 % is worth mentioning.

3.7 SoC Estimation Accuracy

The SoC estimation is essential in non-road vehicles, since the occurring high power densities cause measurement inaccuracies of the on-board sensors. Thus, large SoC mismatches are observable during operation in case of a current-accumulated SoC. Several initializations are required during operation to increase the accuracy, but mostly non-road vehicles are operated without breaks due to which a SoC reset based on standby conditions is not possible. For the SoC validation signal of the available real battery module, in Fig. 3.18, the comparison is depicted between the current-accumulated SoC and the implemented BMS SoC. At this point, it is important to mention that a high accurate current sensor implemented in the battery tester is used to measure the current signal, which admits the comparison.

The first, second, and third subplots in Fig. 3.18 show the voltage, current, and SoC signal, respectively. In the last subplot, the difference between the current-accumulated SoC and the BMS-provided SoC is depicted, where a significant mis-match is clearly observable. Note that the SoC validation signal is not measured in one test run, due to which the observable SoC drift of the real BMS is cleared after the cycles without superimposed constant current. Based on this information, the assumption of inaccurate SoC estimation of the BMS is shown to be true.

In the following, the results achieved with the LMN-based SoC estimation for the battery module are presented. Since cell monitoring is important in terms of safety issues of battery modules, the SoC estimation methodology is also applicable to battery cells. To this end, SoC estimation results obtained for the battery cells are presented additionally. Nevertheless, an overall SoC estimation is desired, for which the module SoC estimation is of main interest.

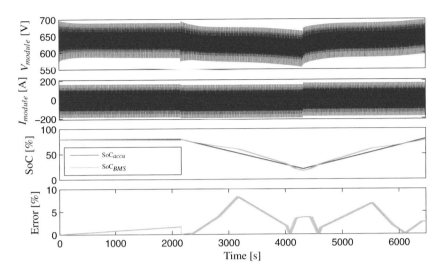

Fig. 3.18 SoC estimation accuracy shown on a SoC validation signal measured at a real battery module

3.7.1 Battery Module SoC Estimation Results

The battery module model Σ_4 is used in the fuzzy observer to estimate the SoC of the battery module. As discussed in Sect. 2.6.2, estimation accuracy of the Kalman filters can be tuned by choosing the corresponding process $\mathbf{Q} = \mathbf{1} \cdot Q$ and measurement $\mathbf{R} = \mathbf{1} \cdot R$ noise variance matrices, where $\mathbf{1} \in \mathbf{R}^{n \times n}$ indicates a one matrix with corresponding dimension n. For the given battery module, a measurement noise variance is chosen to $R = 25000$, while the process noise variance is chosen between $Q = 0.01$ and $Q = 1$ to show the influence. The results are depicted in Fig. 3.19.

The upper subplot in Fig. 3.19 shows the estimated SoC, while in the lower sub-plot, the difference to the current-accumulated SoC is depicted. It can be seen that the accuracy increases with smaller process noise, which in other terms can be interpreted as larger weighting of the current accumulation within the augmented observer model. The disadvantage of a small Q is, though, that the convergence in case of a wrong initial SoC is significantly slower. Since a correctly approximated SoC value is usually obtained after standby, the OCV-based SoC estimation is accurate enough to initialize the fuzzy observer, and the process noise can be chosen small. Nevertheless, the convergence speed of the filter is also important, and the performance of the observer in terms of convergence should be discussed in the following. To this end, unrealistic initial SoC values are used to test the convergence speed of the filter with different process noise variances. In Fig. 3.20, the results for initial SoC at $SoC_{init} = 0\,\%$ and $SoC_{init} = 200\,\%$ are depicted. Note that $SoC_{init} = 200\,\%$ is not feasible in general, but shows that the filter converges from any initial SoC.

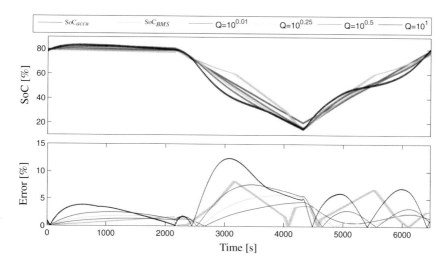

Fig. 3.19 Accuracy of LMN-based SoC estimation on real battery module

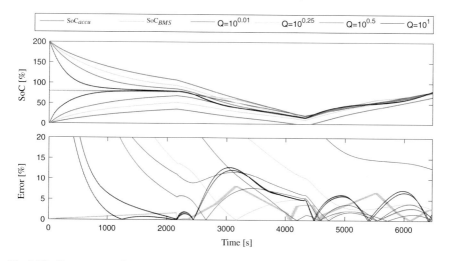

Fig. 3.20 Convergence of LMN-based SoC estimation on real battery module

As can be seen in the second subplot, the larger choice of the process noise variance has a positive influence on the convergence speed, but the accuracy at converged state is poor. Therefore, a trade-off between a fast convergence and SoC estimation accuracy needs to be made, and the best application-specific alternative must be chosen. In comparison to the SoC value obtained through the BMS of the module, the differences are low due to the precise LMN-based battery model. Nevertheless, small process noise leads to unsufficient performance, since the actual SoC of the module cannot be reached if the initial SoC is too far from the actual value.

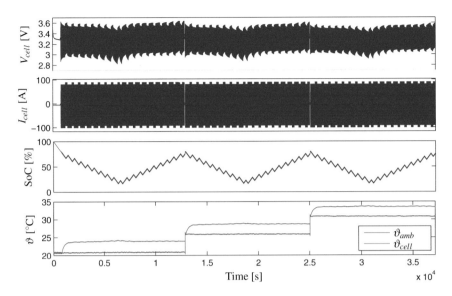

Fig. 3.21 SoC estimation accuracy shown on the SoC validation signal at different temperatures measured on battery Cell **B**

3.7.2 Battery Cell SoC Estimation Results

Since monitoring purposes of the BMS are also important, in the following, the SoC estimation applied to battery cells is reviewed shortly. The BMS requires the monitoring of single-cell voltages and conditions, which provides a possibility to estimate the cell SoC beside of the module SoC simultaneously. Note that the estimation of the cell SoC is more difficult than the module SoC, since an already very small voltage difference indicates a large SoC difference. This is caused by the flat discharge voltage behavior as discussed in Sect. 2.2.3.1 and Fig. 2.5. Nevertheless, similar to the battery module, the filter parameters can be chosen correspondingly to the desired filter properties.

Based on the merged SoC validation signals measured at different temperatures, the SoC estimation of battery cells should be presented. In Fig. 3.21, the validation data is depicted. The merging of the individual data sets can be observed clearly in the temperature and SoC signal, respectively, where steps in the ambient temperature and SoC due to a SoC reset occur. In Fig. 3.22, the battery cell SoC estimation results for the initial SoC values $SoC_{init} = 0\%$, $SoC_{init} = SoC$, and $SoC_{init} = 200\%$ are depicted, which are obtained for a chosen measurement noise variance of $R = 1 \cdot 25000$ and a set process noise variance of $Q = 1 \times 10^{0.01}$.

The convergence of the filter is clearly observable in Fig. 3.22, where at the end of the signal, all SoC signals are almost identical. Although the initial condition of $SoC_{init} = 200\%$ is physically impossible, the filter is able to converge to the correct SoC. As can be seen in Subplot two of Fig. 3.22, the accuracy for the first SoC

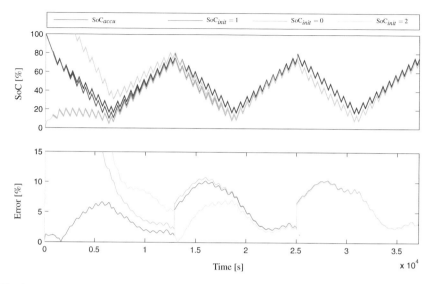

Fig. 3.22 SoC estimation of battery cells with different initial SoC values

validation signal is better in comparison to the others, although a SoC estimation error is observable for the discharge part of the signal. This is caused by the model mismatch, although the accuracy of the LMN battery cell model is high. For the higher temperature validation signals, the error is significantly higher, because the SoC is reset in the reference SoC integration. In combination with the model mismatch, this leads to an estimation error of up to 10 %. Nevertheless, the voltage behavior is very similar around a wide range of approximately 50 % SoC, which makes an accurate estimation significantly below 10 % very difficult. In comparison with the battery module SoC estimation, this influence is reduced due to the higher voltage deviations.

Chapter 4
Energy Management

Abstract Energy management in hybrid electric vehicles in non-road application is the superordinated control of the overall energy flows and influences the system behavior of the powertrain significantly. Minimization of the overall energy flows, exhaust emissions as well as fuel consumption are the primary objectives of the EMS, while all physical constraints of the system need to be taken into account (Sciarretta and Guzzella 2007). Due to the multiple energy conversions from and into the battery, the minimization of the losses is essential and can only be achieved by an optimization of the overall system. However, the overall system is strongly nonlinear in general and differs significantly between different applications. A generic EMS in terms of applicability and cost reductions is therefore favorable. In this chapter, the energy management system is described in detail and methodologies to improve the EMS performance are introduced.

Keywords Non-road hybrid electric vehicle (HEV) · Nonlinear model predictive control · Load and cycle prediction · Stability · Convergence · Real-time control

4.1 Introduction

4.1.1 Challenges for Energy Management Systems

The unknown high dynamic load acting on the powertrain is characteristic for non-road vehicles. Compared to on-road vehicles, the driving directions and conditions as well as working tasks change very often in non-road vehicles and depend directly on the driver. In this context, the control of the powertrain has to cope with these circumstances to fulfill the requirement of a robust operation and the obviation of engine stalling. Additional to the external circumstances, different strategies such as a limited ICE torque gradient to reduce exhaust emissions (phlegmatization) (Lindenkamp and Tilch 2012; Nüesch et al. 2014) and a lowered average rotational speed to reduce fuel consumption (downspeeding) influence the degrees of freedom to control the powertrain.

© The Author(s) 2016
J. Unger et al., *Energy Efficient Non-Road Hybrid Electric Vehicles*,
SpringerBriefs in Applied Sciences and Technology,
DOI 10.1007/978-3-319-29796-5_4

Vehicle misuse or driver errors often lead to unexpected load peaks, while noise corrupted (measured) variables pretend the wrong actual state of the vehicle. Especially the actual load value, which is obtained by the calculation from different signals, and the torque values, which are obtained by the control units instead of being measured, are relevant for the controller. In terms of minimization of the losses, cyclically operated vehicles are often especially challenging, because higher SoC variations increase the usage of the electrical system. Nevertheless, the potential for fuel and emission reduction of cyclical operated vehicles is higher if the load cycle can be considered in the EMS. The required predictions of the future load trajectories are difficult to achieve, though, and can only be based on statistical evidence, which is subject to inaccuracies. Driver information, which is mostly not available in advance, can be considered for prediction of future load requirements. However, the difficulty is to predict the future load cycle to a sufficiently accurate extent and in particularly at often changing load cycles. In the end, a multidimensional nonlinear optimization problem results that needs to be solved in real time in each time step to obtain the optimal control values, which is additionally tightened by the nonlinear behavior of the electrical system (Lyshevski 2000).

4.1.2   State-of-the-Art

The field of EMS in HEV is strongly investigated, while only some of the authors consider load predictions to improve their proposed energy management systems. In the following, the state-of-the-art in EMS and load prediction is reviewed.

Desai and Williamson (2009) classified and compared various control strategies to provide novel development directions in HEV. In Pisu and Rizzoni (2007), three different energy management approaches are compared, but a-priori knowledge of the future load demand is not considered. A model-based strategy to control the load of an on-road parallel HEV is proposed by Sciarretta et al. (2004), which also does not include the future driving conditions in the concept. In Lin et al. (2003), dynamic programming is applied to simulation data in order to extract a rule-based controller as power management for a parallel hybrid truck. By accepting a small increase in fuel consumption, real-time capability and a significant emission reduction could be achieved by the extracted controller. Poursamad and Montazeri (2008) tuned a genetic-fuzzy control strategy with a genetic algorithm based on three driving cycles including NEDC,[1] FTP[2] and TEH-CAR.[3] Stochastic dynamic programming is used by Moura et al. (2011) to optimize a power management for plug-in HEV over a distribution of drive cycles rather than a single cycle. Furthermore, fuel and electricity usage are explicitly traded off, and the impact of variations in relative fuel-to-electricity pricing is considered. In general, offline optimization may lead

---

[1]New European Driving Cycle.

[2]EPA Federal Test Procedure.

[3]Car driving cycle for the capital city of Tehran.

to suboptimal control of non-road vehicles, since the load demand is unknown in advance.

For on-road vehicles, several approaches are known that predict the future load demand in order to be considered in the energy management. An approach with exponentially decreasing torque demand across the prediction horizon is used by Yan et al. (2012) and Borhan et al. (2010). Lin et al. (2004) proposed a driving pattern recognition to classify the current state into one of the six representative driving patterns, for which implementable, sub-optimal controllers are extracted. Analytical approaches for future load demands are very difficult or impossible to find, since load trajectories of non-road vehicles are normally unknown in advance (Unger et al. 2015). A genetic-fuzzy HEV control is used by Montazeri-Gh et al. (2008), which classifies driving patterns by Hidden Markov models. Hulnhagen et al. (2010) used a probabilistic finite-state machine to merge basic maneuver elements to a driving pattern of on-road vehicles. The classification into driving patterns or basic maneuver elements is not suitable for non-road HEV, because in most non-road applications the elements are not clearly assignable. Since map information is not available for non-road vehicles, GPS data-based path-forecasting for trajectory planning (c.f. Katsargyri et al. 2009, Ganji and Kouzani 2011), as mentioned for on-road vehicles by Sciarretta and Guzzella (2007) and Back (2005), is not feasible. Payri et al. (2014) extract an estimate for future driving conditions by analyzing the power demands in a given receding horizon and use the information in a stochastic controller. The prediction of recurrent load cycles used by Mayr et al. (2011a) is based on the cross correlation function algorithm proposed by Lorenz and Kozek (2007) that originally detected cycle boundaries automatically for a statistic evaluation. The disadvantage of the autocorrelation function is the slow detection of a changing load cycle at significantly different cycle times (Unger et al. 2015).

In summary, no online implemented MPC for non-road application is known that features real-time prediction of the future load demand in the MPC.

4.1.3 Solution Approach

In the following, a cascaded control concept is proposed for the parallel hybrid powertrain schematically depicted in Fig. 1.4. Additionally, two methodologies are presented to predict the future short-term load and to detect recurrent load cycles in the past load signal. The proposed cascaded control concept consists of two separate controllers, which are both designed as model predictive controllers. In this context, MPCs provide the required possibilities to achieve an optimization of the overall system with consideration of the load predictions (Mayne et al. 2000). Note that an advanced solving algorithm as presented for the optimization problem and sufficient computational capabilities in the HCU are required to achieve real-time capability.

The operation of the powertrain on the optimal load cycle trajectory can increase the overall efficiency and fully exploit the battery capacity. To this end, the cross correlation function (CCF) is used by a cycle detection algorithm to identify a recurrent load cycle within the past load signal that is further used to build the future load trajectory for the master MPC. Based on the detected recurrent load cycle, the master MPC is designed to provide the optimal operation point of the ICE as demand value for the slave controller, while the SoC of the battery is kept at the demand SoC, and constraints as well as the nonlinearities of the electrical system are considered. In this context, the electrical system of the hybrid powertrain provides for a boost at high load demands and to recuperate regenerative loads, but the battery capacity and physical constraints restrict its usage, which furthermore reduces the degrees of freedom to a certain operation range of the ICE. A nonlinear optimization problem results within the master controller, which is iteratively solved by a real-time capable relaxation approach.

Purpose of the slave MPC is to apply the demands of the master MPC to the powertrain, while constraints are considered. The ICE dynamics are significantly limited by phlegmatization and downspeeding strategies, which is disadvantageous for the high dynamic requirements of non-road vehicles. However, in order to prevent engine stalling and to be able to lower the average rotational speed of the powertrain, Bayesian inference is applied to statistically predict the future short-term load by means of available vehicle information such as accelerator position or driving speed (load prediction algorithm). This approach provides insight into the intentions of the driver and allows an increase of the rotational speed in case of sudden load peaks, which compensates the disadvantage of the restricted ICE dynamics. Nevertheless, the short-term load prediction is based on probabilities and misprediction is considered. Furthermore, the convergence of the master MPC and the essential stability of the slave MPC are discussed, since the controller directly acts on the vehicle.

4.2 Basic Concept of Model Predictive Control

Model predictive control is an advanced method in control theory, which is suited to solve constrained control problems in the time domain. Different types such as Dynamic Matrix Control (DMC), Model Algorithmic Control (MAC), Generalized Predictive Control (GPC), etc. are known, but the main principle is similar for all approaches. The idea of the concept is to obtain the control moves for a process based on an online optimization of an objective (cost) function. A dynamic process model is used to predict the system output over a so-called *prediction horizon*. In order to minimize the control error, the sequence of control moves over a so-called *control horizon* is optimized until the cost function reaches a minimum. Only the first control move in the obtained sequence is applied to the plant, the horizons are moved for one sample (*receding horizon principle*), and the optimization is done anew.

Process Model. The process model has a substantial influence on the MPC. It is therefore important that the process model fully captures the process dynamics and allows accurate predictions of the output. In general, two main methodologies can be used to obtain a process model: Modeling by physical principles to obtain a *white box* model or data-based system identification to obtain a *black box* model. Independent to the method of parameter determination, different types of process model representations can be used in the MPC. Commonly, a state space representation

$$\mathbf{x}(k+1) = \mathbf{A}\mathbf{x}(k) + \mathbf{B}\mathbf{u}(k) + \mathbf{E}\mathbf{z}(k), \tag{4.1}$$

$$y(k) = \mathbf{C}\mathbf{x}(k), \tag{4.2}$$

is implemented, though, where $\mathbf{A}, \mathbf{B}, \mathbf{C}, \mathbf{D}, \mathbf{E}$ are the system matrices including the system dynamics, $\mathbf{x}$ is the state vector, $\mathbf{u}$ is the input vector, y is the output, $\mathbf{z}$ is the disturbance vector, and the time instant is denoted by k.

In many cases, real processes are nonlinear, which means that system parameters depend on system states or/and time. Linearization is one possibility to map the system behavior into one invariant SS-system with the disadvantage of reduced model accuracy. Nonlinear system behavior can be represented by, e.g., local model networks (see Sect. 2.2), which use, in principle, local linearizations to calculate the nonlinear system behavior. A fuzzy MPC (Babuska 1998) can handle the LMN model to consider the nonlinearities. Another nonlinear MPC approach is presented in the following sections, and more information can be found in e.g. Allgöwer and Zheng (2000).

In Fig. 4.1, the principle of MPC is depicted for a single-input-single-output (SISO) system. The blue trajectory indicates the system output if the red control moves and the gray disturbance is acting on the plant, while the green trajectory shows the system states. As can be seen in the figure, the input and state trajectory keep the applied constraints of the plant.

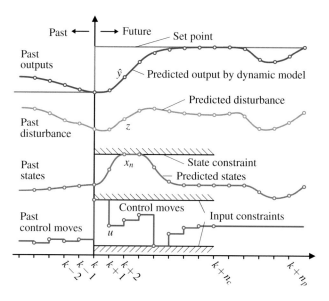

Fig. 4.1 Principle of model predictive control trajectories

Prediction and Control Horizon. As can further be seen in Fig. 4.1, the control moves are kept constant on the last calculated value $u(k + i) = u(k + n_c)$ for $i = n_c, \ldots, n_p - 1$ until the end of the prediction horizon n_p (Babuska 1998; Wang 2009). In order to predict further into the future, the prediction horizon n_p is chosen larger than the control horizon n_c, since n_c determines the number of manipulated variables and thus affects the computation time. Due to the process model, the predicted output values $\hat{y}(k + i | k)$, $i = 1, \ldots, n_p$ are influenced by the current state of the process at time instant k and the future control moves $u(k + i)$, $i = 0, \ldots, n_c - 1$. However, the prediction and control horizons are tuning parameters, and no general rule exists for optimal horizons.

Objective Function. In order to optimize the control moves, an objective function needs to be formulated. A common cost function J is

$$J = \sum_{i=1}^{n_p} Q_i [r(k+i) - \hat{y}(k+i|k)]^2 + \sum_{i=1}^{n_c} \Delta u^T (k+i-1) R_i \Delta u(k+i-1), \qquad (4.3)$$

where n_p and n_c denote the prediction and control horizon, respectively, the prediction of the output is indicated by $\hat{y}(k + i|k), i = 1, 2, \ldots$ for time instant k at prediction instant i and Q_i, R_i weight output error and control effort, respectively. Note that in absence of constraints, an explicit analytical solution can be found due to the quadratic nature of the cost function.

4.3   Cascaded Model Predictive Controller Design

In this section, the cascaded model predictive controller design for parallel hybrid electric vehicles is discussed. The architecture and system models are developed first, before constraints and both MPCs are discussed in detail. Methodologies to predict the future load trajectory are discussed in the next section (see Sect. 4.4).

4.3.1   Architecture of the Control Concept

In the following, the controller architecture is briefly discussed to give a better overview of the concept. The powertrain consists of significantly different dynamics. Rotational speed and the torques represent fast dynamics within the ms time range. On the other hand, the battery's state of charge is only slowly changing within the seconds time range. This difference is addressed by the cascaded control concept, which includes a slow master controller and a fast slave controller both established as MPC. However, note that the interaction between INV and battery also has a fast electrical impedance time constant, which is not relevant to the change in SoC referred to in this book (Hametner et al. 2014).

In Fig. 4.2, the concept is depicted schematically, where the light blue area represents the EMS and the white area the powertrain. The load P_{load} (red) acts as the unknown disturbance of the powertrain, n_{GW} is the driving speed, and α is the accelerator angle that is directly influenced by the driver. Note that the actual SoC can be obtained by the SoC estimator presented in Sect. 2.6 and is therefore assumed to be known. The aim of the master MPC is to hold the control variable SoC at the demand value SoC_{dmd}, while a full recurrent load cycle as well as all constraints are considered. In the concept, the cycle detection (Cycle Det.) provides the predicted future load trajectory $\mathbf{Z}_{CD}$ for the master MPC, while due to an optimization, the optimal values for the demand rotational speed ω_{dmd} and ICE torque $T_{ice,dmd}$ are obtained. Note that the prediction of the SoC trajectory within the master MPC depends on the battery conditions and is thus strongly nonlinear. In principle, the output of the master MPC defines the operation point of the ICE. The aim of the slave MPC is then to compensate unpredicted short-term load peaks by considering the load prediction (Load Pred.) methodology that provides the short-term load trajectory $\mathbf{Z}_{LP}$ for the slave MPC. Constraints are especially important to be kept by the slave MPC, since the manipulated variables ICE torque $T_{ice,set}$ and ISG torque $T_{isg,set}$ directly act on the vehicle. In this context, it is worth mentioning that the rotational speed ω is only controlled by ICE torque T_{ice} and ISG torque T_{isg}, respectively. Note that sufficient powertrain dynamics in case of strong load gradients must be guaranteed, which due to the slow sampling time $t_{s,m} = 0.25\,\text{s}$ of the master MPC must be compensated by the slave MPC. To this end, undetected or wrongly detected load cycles must be compensated. The CD and LP methodologies aim to find the a-priori unknown future load trajectories only based on the system states P_{load}, α and n_{GW} actually available during operation, which is discussed in more detail in the next subsection (cf. Sect. 4.4).

4.3.2 System Models for Controller Design

The models for the slow as well as fast system dynamics of the plant are required by the design of the MPCs and are developed in the following.

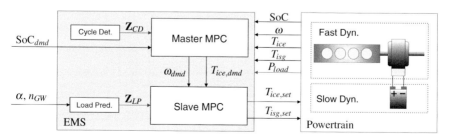

Fig. 4.2 Schematic workflow of the cascaded control concept

First order lag elements and the principle of angular momentum, respectively, can be used to model the fast dynamics of the rotational speed and the torques in the continuous time domain (c.f. Powell 1979, Fekri and Assadian 2012), which follow to

$$T_{ice}(t) + T_{isg}(t) - T_{load}(t) = \Theta \, \dot{\omega}(t), \tag{4.4}$$

$$\tau_{ice} \, \dot{T}_{ice}(t) + T_{ice}(t) = T_{ice,set}(t), \tag{4.5}$$

$$\tau_{isg} \, \dot{T}_{isg}(t) + T_{isg}(t) = T_{isg,set}(t). \tag{4.6}$$

In Eqs. (4.4)–(4.6), the powertrain's total moment of inertia is denoted by Θ, the time constants of the torques are denoted by τ_{ice}, τ_{isg} and $T_{load}(t) = P_{load}(t)/\omega(t)$. The ICE is in general a nonlinear system, but the dependence of T_{ice} on ω is neglected. In order to consider the variant time constant of the ICE, τ_{ice} can be updated in each time instant to improve the control quality of the slave controller. Nevertheless, for the main focus of this book, the simplified model of the ICE dynamic is satisfactory, because it is important that time delays from the components are addressed in the process model of the controller and feasibility of the concept can be shown (Unger et al. 2015).

For the SS-representation of the slave MPC process model, the input vector $\mathbf{u}_s$, state vector $\mathbf{x}_{d,s}$, output $y_{d,s}$ and disturbance z_s can be chosen to

$$\mathbf{x}_{d,s} = \begin{bmatrix} \omega \\ T_{ice} \\ T_{isg} \end{bmatrix}, \ y_{d,s} = \omega, \ \mathbf{u}_s = \begin{bmatrix} T_{ice,set} \\ T_{isg,set} \end{bmatrix}, \ z_s = T_{load}, \tag{4.7}$$

while the fast sampling time of the slave controller is defined by $t_{s,s} = 10\,\text{ms}$. The discrete linear state space slave controller model is obtained by

$$\mathbf{x}_{d,s}(k+1) = \mathbf{A}_{d,s}\mathbf{x}_{d,s}(k) + \mathbf{B}_{d,s}\mathbf{u}_s(k) + \mathbf{E}_{d,s}T_{load}(k), \tag{4.8}$$

$$\omega(k) = \mathbf{C}_{d,s}\mathbf{x}_{d,s}(k), \tag{4.9}$$

where

$$\mathbf{A}_{d,s} = \begin{bmatrix} 1 & \frac{t_{s,s}}{\Theta} & \frac{t_{s,s}}{\Theta} \\ 0 & \frac{\tau_{ice}}{\tau_{ice}+t_{s,s}} & 0 \\ 0 & 0 & \frac{\tau_{isg}}{\tau_{isg}+t_{s,s}} \end{bmatrix}, \ \mathbf{C}_{d,s} = \begin{bmatrix} 1 & 0 & 0 \end{bmatrix},$$

$$\mathbf{B}_{d,s} = \begin{bmatrix} 0 & 0 \\ \frac{t_{s,s}}{\tau_{ice}+t_{s,s}} & 0 \\ 0 & \frac{t_{s,s}}{\tau_{isg}+t_{s,s}} \end{bmatrix}, \ \mathbf{E}_{d,s} = \begin{bmatrix} \frac{t_{s,s}}{\Theta} \\ 0 \\ 0 \end{bmatrix}. \tag{4.10}$$

The process model for the master MPC needs to constitute the relation between ISG torque and the resulting battery SoC (Unger et al. 2015). Based on the fact that integral of the battery current $I(t)$ and SoC are proportional (Hametner and Jakubek 2013), a corresponding model is obtained by

$$\text{SoC}(t) = \text{SoC}_{init} + \frac{1}{Q_{c,batt}} \int_0^t \eta_{batt,Cou}(I(v)) \, I(v) dv, \qquad (4.11)$$

$$I(t) = k_I\left(T_{isg,dmd}, \omega_{dmd}, V\right) T_{isg,dmd}(t), \qquad (4.12)$$

where SoC_{init} is the initial SoC of the battery. The relation between current and SoC is defined by battery capacity $Q_{c,batt}$ and coulombic efficiency $\eta_{batt,Cou}$, while the proportionality between ISG torque and current is denoted by k_I. Note that the battery current $I(t)$ is a nonlinear function of ISG torque, speed and battery voltage. Since the voltage depends on the current itself, the equations lead to an implicit form (Mayr et al. 2011a).

The difference between load and ICE torque defines the ISG torque, which follows to

$$T_{isg,dmd}(t) = \frac{P_{load}(t)}{\omega_{dmd}(t)} - T_{ice,dmd}(t). \qquad (4.13)$$

The nonlinear proportionality k_I is invariant during transient operation and can therefore be statically determined by test bed measurements. A characteristic map of k_I can be extracted, which includes the motor and drive efficiencies at different voltage levels. By means of analytical polynomial surface approximation of each voltage level of the identified characteristic map, the current can be expressed by

$$\begin{aligned} I(\omega_{dmd}, T_{isg})_v = {} & p_{00,v} + p_{10,v} \, \omega_{dmd} + p_{01,v} \, T_{isg} \\ & + p_{20,v} \, \omega_{dmd}^2 + p_{02,v} \, T_{isg}^2 + p_{11,v} \, \omega_{dmd} T_{isg}, \end{aligned} \qquad (4.14)$$

where p_{ij} are least squares identified parameters and v corresponds to the voltage level. Note that the analytical approximation has sufficient accuracy and simplifies the further process significantly. Differentiation of $I(\omega_{dmd}, T_{isg})_v$ with respect to the ISG torque follows to

$$k_{I,v} = \frac{dI(\omega_{dmd}, T_{ISG})_v}{dT_{isg}} = p_{01,v} + 2\, p_{02,v} \, T_{isg} + p_{11,v} \, \omega_{dmd}. \qquad (4.15)$$

In order to obtain the value of k_I, the obtained analytic equation for $k_{I,v}$ can be linearly interpolated between the different voltage levels

$$k_I = k_{I,v_i} + \frac{k_{I,v_j} - k_{I,v_i}}{v_j - v_i}\,(V - v_i), \qquad (4.16)$$

where V is the terminal voltage of the battery. Note that for the purpose of the state prediction, a battery model is required to provide a value for V in the prediction.

The system behavior of electrochemical batteries is nonlinearly dependent on SoC, temperature ϑ_{batt} and current I (Unger et al. 2014). Nonlinear effects such as relaxation and hysteresis are observable as well (Plett 2004b). A powerful approach are local model networks, though several approaches are known for battery modeling (see Chap. 2). LMN comprise of local linear dynamic models, each of which is valid in a certain operating region of a partition space (Hametner et al. 2012). The mentioned effects are considered by corresponding inputs, while the global nonlinear model output is obtained by weighted aggregation of the LLM outputs (Hametner et al. 2013b). Automatic iterative algorithms (e.g., Nelles and Isermann 1996, Jakubek and Hametner 2009) are used to built the LMN structure. A complete discussion about battery modeling with LMN is given in Chap. 2.

Since dynamic models require past system inputs and outputs, the output trajectory for a given input trajectory requires a simulation. The computational demand of dynamic models is therefore significantly higher than for parameter varying static models. Due to this reason, in the EMS, a simplified equivalent circuit battery model is used, which is depicted schematically in Fig. 4.3 (see e.g. Hu et al. 2009b).

In order to consider the nonlinear parameter varying nature of batteries, look-up tables of the values of inner resistance R_{int} (SoC, ϑ_{batt}, I) and open circuit voltage V_0 (SoC, ϑ_{batt}) are extracted from an identified LMN. Applying Kirchhoff's second law, the battery voltage follows by

$$V(t) = V_0 (\text{SoC}, \vartheta_{batt}) - R_{int} (\text{SoC}, \vartheta_{batt}, I) I(t). \qquad (4.17)$$

Note that other modeling and prediction inaccuracies have large influences due to which the effect of the decreased battery model accuracy is of no consequence (Unger et al. 2015). Discretizing Eq. (4.11) using the master sampling time $t_{s,m}$ leads to a discrete time state space representation of the SoC model. Accumulation of the discrete current $I(k)$ multiplied with

$$k_{batt}(k) = \frac{\eta_{batt,Cou}(k) \, t_{s,m}}{Q_{c,batt}}, \qquad (4.18)$$

Fig. 4.3 Schematic overview of the used ECM battery model

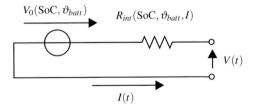

where the coulombic efficiency $\eta_{batt,Cou}(k)$ of the battery

$$\eta_{batt,Cou}(k) = \begin{cases} \eta_{cha} = 0.99 & \text{for } I(k) > 0, \\ \eta_{dis} = 1 & \text{for } I(k) < 0, \end{cases} \qquad (4.19)$$

is obtained by Verbrugge and Tate (2004), approximates the integral in Eq. (4.11). Choosing the input u_m, state $x_{d,m}$, output $y_{d,m}$ and disturbance z_m of the system by

$$x_{d,m} = y_{d,m} = \text{SoC}, \quad u_m = T_{ice,dmd}, \quad z_m = T_{load}, \qquad (4.20)$$

the discrete SS representation of Eqs. (4.11)–(4.19) follows to

$$x_{d,m}(k+1) = A_{d,m}x_{d,m}(k) + B_{d,m}(k)u_m(k) + E_{d,m}(k)z_m(k), \qquad (4.21)$$
$$y_{d,m}(k) = C_{d,m}x_{d,m}(k), \qquad (4.22)$$

with

$$A_{d,m} = 1, \ B_{d,m}(k) = -k_I(x_{d,m}, u_m, z_m, k)\, k_{batt}(u_m, z_m, k),$$
$$C_{d,m} = 1, \ E_{d,m}(k) = k_I(x_{d,m}, u_m, z_m, k)\, k_{batt}(u_m, z_m, k), \qquad (4.23)$$

where the parameter varying structure of the ECM is directly integrated in k_I.

4.3.3 Structured Constraints for Controllers

The system states of both controllers do not directly include the battery current and voltage as well as the temperature of ISG and battery. Thus, these constraints need to be considered indirectly by structured constraints. In this context, structured constraints means to apply only the most constraining value, while current, voltage, and temperatures can be limited by increasing/decreasing the ISG torque. Equations (4.24)–(4.32) summarize all relevant structured constraints:

$$\hat{\omega}_{min} = \max \begin{cases} \omega_{min} \\ \omega_{min,driver} \end{cases}, \qquad (4.24)$$

$$\hat{T}_{ice,min} = \max \begin{cases} T_{ice,min}(\omega) \\ \frac{P_{load}}{\omega_{dmd}} - \hat{T}_{isg,min} \end{cases}, \qquad (4.25)$$

$$\hat{T}_{ice,max} = \min \begin{cases} T_{ice,max}(\omega) \\ \frac{P_{load}}{\omega_{dmd}} - \hat{T}_{isg,max} \end{cases}, \qquad (4.26)$$

$$\hat{T}_{isg,min} = \max \begin{cases} T_{isg,min}(\omega, V) \\ T_{isg,min}(\vartheta_{isg}, \vartheta_{batt}) \\ T_{isg,min}(\text{SoC}) \\ T_{isg,min}(V, I) \end{cases} , \qquad (4.27)$$

$$\hat{T}_{isg,max} = \min \begin{cases} T_{isg,max}(\omega, V) \\ T_{isg,max}(\vartheta_{isg}, \vartheta_{batt}) \\ T_{isg,max}(\text{SoC}) \\ T_{isg,max}(V, I) \end{cases} , \qquad (4.28)$$

$$I_{min} \le I \le I_{max}, \qquad (4.29)$$
$$V_{min} \le V \le V_{max}, \qquad (4.30)$$
$$\vartheta_{batt,min} \le \vartheta_{batt} \le \vartheta_{batt,max}, \qquad (4.31)$$
$$\vartheta_{isg,min} \le \vartheta_{isg} \le \vartheta_{isg,max}, \qquad (4.32)$$

where $\omega_{min,driver}$ is the minimal rotational speed required by the driver and ϑ denotes the temperatures. Using the actual states of the system, the characteristic maps provide the minimal (min) and maximal (max) system constraints (see corresponding function arguments). Note that $f(V, I)$ additionally constrains the maximal/minimal battery power.

The finally applied set of constraints for the slave controller can be summarized by

$$\mathcal{C}_s = \begin{cases} \hat{\omega}_{min} \le \omega \le \omega_{max} \\ \Delta T_{ice,min} \le \Delta T_{ice,set} \le \Delta T_{ice,max} \\ \Delta T_{isg,min} \le \Delta T_{isg,set} \le \Delta T_{isg,max} \\ \hat{T}_{ice,min} \le T_{ice,set} \le \hat{T}_{ice,max} \\ \hat{T}_{isg,min} \le T_{isg,set} \le \hat{T}_{isg,max} \end{cases} , \qquad (4.33)$$

where the backwards difference operator Δ indicates rate constraints. The set of constraints for the master controller follows to

$$\mathcal{C}_m = \begin{cases} \Delta T_{ice,min} \le \Delta T_{ice,dmd} \le \Delta T_{ice,max} \\ \hat{\omega}_{min} \le \omega_{dmd} \le \omega_{max} \\ \hat{T}_{ice,min} \le T_{ice,dmd} \le \hat{T}_{ice,max} \\ \text{SoC}_{min} \le \text{SoC} \le \text{SoC}_{max} \end{cases} . \qquad (4.34)$$

All given constraints are considered in the control system.

4.3.4 Slave Controller

In the following, the slave MPC is discussed in detail. The demand values rotational speed ω_{dmd} and ICE torque $T_{ice,dmd}$ are applied to the plant, while the used manipulated variables $T_{ice,set}$, $T_{isg,set}$ need to keep the constraints.

4.3.4.1 Slave MPC Formulation

An augmentation of the process model Eqs. (4.8) and (4.9) provides the possibility to directly constrain the gradient of the manipulated variables as well as allows an offset free control by avoiding a steady state bias (Camacho and Bordons 1999). The augmentation is done by embedding an integrator, which leads to an incremental plant description (Ogata 1995)

$$
\underbrace{\begin{bmatrix} \Delta \mathbf{x}_{d,s}(k+1) \\ y_{d,s}(k+1) \end{bmatrix}}_{\mathbf{x}_s(k+1)} = \underbrace{\begin{bmatrix} \mathbf{A}_{d,s} & \mathbf{0} \\ \mathbf{C}_{d,s}\mathbf{A}_{d,s} & 1 \end{bmatrix}}_{\mathbf{A}_s} \underbrace{\begin{bmatrix} \Delta \mathbf{x}_{d,s}(k) \\ y_{d,s}(k) \end{bmatrix}}_{\mathbf{x}_s(k)}
$$
$$
+ \underbrace{\begin{bmatrix} \mathbf{B}_{d,s} \\ \mathbf{C}_{d,s}\mathbf{B}_{d,s} \end{bmatrix}}_{\mathbf{B}_s} \Delta \mathbf{u}_s(k) + \underbrace{\begin{bmatrix} \mathbf{E}_{d,s} \\ \mathbf{C}_{d,s}\mathbf{E}_{d,s} \end{bmatrix}}_{\mathbf{E}_s} \Delta z_s(k), \qquad (4.35)
$$

$$
\omega(k) = \underbrace{\begin{bmatrix} \mathbf{0} & 1 \end{bmatrix}}_{\mathbf{C}_s} \mathbf{x}_s(k), \qquad (4.36)
$$

where $\mathbf{x}_s$ is the augmented state vector, $\Delta \mathbf{u}_s$ is the incremental input, Δz_s is the incremental disturbance, and $\mathbf{0}$ represents a zero matrix with corresponding dimension. A sequence of $n_c = 25$ incremental control moves

$$
\Delta \mathbf{U}_s(k) = \begin{bmatrix} \Delta \mathbf{u}_s(k+1)^T & \cdots & \Delta \mathbf{u}_s(k+n_c)^T \end{bmatrix}^T, \qquad (4.37)
$$

can be found that minimizes a cost function

$$
J_s = Q_s \sum_{i=1}^{n_p-1} (\omega_{dmd}(k+i) - \omega(k+i))^2 + \sum_{i=0}^{n_c-1} (\Delta \mathbf{u}_s^T(k+i)\mathbf{R}_s \Delta \mathbf{u}_s(k+i))
$$
$$
+ \sum_{i=0}^{n_c-1} (\Delta \tilde{\mathbf{u}}_s^T(k+i)\tilde{\mathbf{R}}_s \Delta \tilde{\mathbf{u}}_s(k+i)) + V_{f,s}(\mathbf{x}_s(k+n_p)), \qquad (4.38)
$$

with $V_{f,s}(\mathbf{x}_s(k+n_p)) = \mathbf{x}_s(k+n_p)^T \mathbf{P}_s \mathbf{x}_s(k+n_p)$, where the term $V_{f,s}$ is a terminal weight. In principle, the cost function J_s has no unique minimum, since the torque split between ICE and ISG is not uniquely defined. For this reason, the term $\Delta \tilde{\mathbf{u}}_s = \begin{bmatrix} T_{ice,dmd} & T_{isg,dmd} \end{bmatrix}^T - \mathbf{u}_s$ penalizes the deviations of the manipulated variables from the demand values provided by the master MPC. Therefore, in case of no active constraints, $\mathbf{u}_s$ is forced to reach the demand values at steady state (c.f. González et al. 2008).

Output, input rate, and input, respectively, are penalized by user-defined, symmetric and positive definite weighting matrices $\mathbf{Q}_s$, $\mathbf{R}_s$ and $\tilde{\mathbf{R}}_s$. The solution of the discrete algebraic Riccati equation (DARE) is used as terminal weight matrix $\mathbf{P}_s$, which follows to

$$\mathbf{P}_s = \mathbf{A}_s^T \mathbf{P}_s \mathbf{A}_s - \mathbf{K}_s^T \left(\mathbf{R}_s + \mathbf{B}_s^T \mathbf{P}_s \mathbf{B}_s \right) \mathbf{K}_s + \mathbf{C}_s^T Q_s \mathbf{C}_s \qquad (4.39)$$

with $\mathbf{K}_s = \left(\mathbf{R}_s + \mathbf{B}_s^T \mathbf{P}_s \mathbf{B}_s \right)^{-1} \mathbf{B}_s^T \mathbf{P}_s \mathbf{A}_s$. Note that, $V_{f,s}$ is used to obtain the stabilizing properties of the linear-quadratic regulator (LQR) (König et al. 2013).

As mentioned in Sect. 4.2, there exists no general rule for the selection of the prediction as well as for the control horizon. However, a guideline for a minimum prediction horizon is to choose it at least large enough to cover the smallest time constant of the system, while a maximum for n_p is given by the maximal system runtime (Camacho and Bordons 1999).

For the example of the wheel loader, usual short-term load peaks occur within 250 ms ahead, which leads to a prediction horizon of $n_p = 25$ chosen in this book. The matrix notation of the output prediction $\hat{\mathbf{Y}}_s$ is obtained by

$$\begin{aligned}
\hat{\mathbf{Y}}_s &= \left[\omega(k+1) \cdots \omega(k+n_p) \; \mathbf{x}_s(k+n_p)^T \right]^T \\
&= \mathbf{F}_s \mathbf{x}_s(k) + \boldsymbol{\Phi}_{s,u} \Delta \mathbf{U}_s(k) + \boldsymbol{\Phi}_{s,z} \Delta \mathbf{Z}_{LP}(k),
\end{aligned} \qquad (4.40)$$

where $\Delta \mathbf{Z}_{LP}$ denotes the incremental disturbance trajectory, which is predicted by the short-term load prediction, and the matrices $\mathbf{F}_s$, $\boldsymbol{\Phi}_{s,u}$ and $\boldsymbol{\Phi}_{s,z}$ predict the system states based on $\mathbf{x}_s(k)$ and $\Delta \mathbf{U}_s(k)$:

$$\mathbf{F}_s = \left[(\mathbf{C}_s \mathbf{A}_s)^T \; \left(\mathbf{C}_s \mathbf{A}_s^2 \right)^T \cdots \left(\mathbf{C}_s \mathbf{A}_s^{n_p-1} \right)^T \; \left(\mathbf{A}_s^{n_p} \right)^T \right]^T, \qquad (4.41)$$

$$\boldsymbol{\Phi}_{s,u} = \begin{bmatrix} \mathbf{C}_s \mathbf{B}_s & \mathbf{0} & \cdots & \mathbf{0} \\ \mathbf{C}_s \mathbf{A}_s \mathbf{B}_s & \mathbf{C}_s \mathbf{B}_s & \cdots & \mathbf{0} \\ \vdots & \vdots & \ddots & \vdots \\ \mathbf{C}_s \mathbf{A}_s^{n_p-2} \mathbf{B}_s & \mathbf{C}_s \mathbf{A}_s^{n_p-3} \mathbf{B}_s & \cdots & \mathbf{0} \\ \mathbf{A}_s^{n_p-1} \mathbf{B}_s & \mathbf{A}_s^{n_p-2} \mathbf{B}_s & \cdots & \mathbf{A}_s^{n_p-n_c} \mathbf{B}_s \end{bmatrix}, \qquad (4.42)$$

$$\boldsymbol{\Phi}_{s,z} = \begin{bmatrix} \mathbf{C}_s \mathbf{E}_s & \mathbf{0} & \cdots & \mathbf{0} \\ \mathbf{C}_s \mathbf{A}_s \mathbf{E}_s & \mathbf{C}_s \mathbf{E}_s & \cdots & \mathbf{0} \\ \vdots & \vdots & \ddots & \vdots \\ \mathbf{C}_s \mathbf{A}_s^{n_p-2} \mathbf{E}_s & \mathbf{C}_s \mathbf{A}_s^{n_p-3} \mathbf{E}_s & \cdots & \mathbf{0} \\ \mathbf{A}_s^{n_p-1} \mathbf{E}_s & \mathbf{A}_s^{n_p-2} \mathbf{E}_s & \cdots & \mathbf{E}_s \end{bmatrix}. \qquad (4.43)$$

Following Wang (2009), the slave controller constraints Eq. (4.33) can be formulated by corresponding linear inequalities and directly implemented in the optimization problem. The compact notation follows to

$$\mathcal{C}_s : \quad \mathbf{M}_s \Delta \mathbf{U}_s \leq \boldsymbol{\gamma}_s,$$

$$\text{with} \quad \mathbf{M}_s = \begin{bmatrix} \mathbf{M}_{s,\Delta u} \\ \mathbf{M}_{s,u} \\ \mathbf{M}_{s,y} \end{bmatrix}, \quad \boldsymbol{\gamma}_s = \begin{bmatrix} \boldsymbol{\gamma}_{s,\Delta u} \\ \boldsymbol{\gamma}_{s,u} \\ \boldsymbol{\gamma}_{s,y} \end{bmatrix}, \tag{4.44}$$

where the indices refer to the rate, input and output constraints, respectively. A formal formulation of the final constrained optimal control problem can be denoted by

$$\mathcal{P}_s(\mathbf{x}_s(k)) : \quad \Delta \mathbf{U}_s^*(\mathbf{x}_s(k)) = \arg \min J_s(\mathbf{x}_s(k), \Delta \mathbf{U}_s)$$

$$\text{s.t. } \mathbf{M}_s \Delta \mathbf{U}_s \leq \boldsymbol{\gamma}_s. \tag{4.45}$$

4.3.4.2 Real-Time Implementation

The given optimization problem Eq. (4.45) needs to be solved in each time instant, which gives a time frame of $t_{s,s} = 10\,\text{ms}$ to solve the full optimization problem. In order to reduce computational loads, in the work of Unger et al. (2012b), an approach is presented to reduce the order of the formulated MPC problem by using principal control moves. Following example, Tondel and Johansen (2002) and Rojas et al. (2004), the idea is to parametrize $\Delta \mathbf{U}_s$ in Eq. (4.37) according to

$$\Delta \mathbf{U}_s = \boldsymbol{\Omega}_s \mathbf{p}_s, \tag{4.46}$$

where $\boldsymbol{\Omega} \in \mathbb{R}^{n_c \cdot n_u \times v_p}$ is a matrix whose columns form a basis for $\Delta \mathbf{U}_s$, n_u is the number of input variables and $\mathbf{p}_s \in \mathbb{R}^{v_p \times 1}$ is the new decision variable with reduced dimension ($v_p \leq n_c \cdot n_u$). The columns of $\boldsymbol{\Omega}$ describe the shape of the control increments of every manipulated variable up to the control horizon and are called *principal control moves*, while the integer variable v_p indicates the number of columns in $\boldsymbol{\Omega}$ and is called the *order of principal control moves* (PCM-order). Replacing $\Delta \mathbf{U}_s$ in the constrained optimization problem Eq. (4.45) by (4.46), the optimization problem Eq. (4.45) with $\mathbf{p}_s$ as the new decision variable follows to

$$\mathcal{P}_p(\mathbf{x}_s(k)) : \quad \mathbf{p}_s^*(\mathbf{x}_s(k)) = \arg \min J_{s,p}(\mathbf{x}_s(k), \mathbf{p}_s) \quad \text{s.t. } \mathbf{M}_s \boldsymbol{\Omega}_s \mathbf{p}_s \leq \boldsymbol{\gamma}_s. \tag{4.47}$$

Another approach for real-time implementation of the MPC is the fast model predictive control (FMPC) algorithm proposed by Wang and Boyd (2010). The particular structure of the MPC problem is exploited to decrease computation times. Since the algorithm is provided for implementation, this method is used, though, the algorithm is appropriately extended to include the disturbance costs in the optimization.

Further approaches are presented in, e.g., Richter et al. (2009), Zeilinger et al. (2011) and Quoc et al. (2012).

4.3.4.3   Stability Analysis Slave Controller

The stability of the slave controller is essential, since it directly acts on the plant. In order to show stability for the constrained controller, distinction is usually made between stability inside a terminal set $\mathbb{X}_f$, where no constraints are active and a region of attraction $\bar{\mathcal{X}}_N$, where constraints are active (Konig et al. 2014; Simon et al. 2012).

Assuming that $\mathbf{x}_s \in \mathbb{X}_f$ and $\tilde{\mathbf{x}}_s(k+1) = (\mathbf{A}_s - \mathbf{B}_s \mathbf{K}_s) \tilde{\mathbf{x}}_s(k) \in \mathbb{X}_f$, a stabilizing control law $\Delta \mathbf{U}_s = -\mathbf{K}_s(\mathbf{x}_s - \mathbf{x}_s^*) = -\mathbf{K}_s \tilde{\mathbf{x}}_s$ inside the positive invariant terminal set $\mathbb{X}_f$ is obtained so that all constraints $\mathcal{C}_s$ are satisfied, when a terminal weight $V_{f,s}$ is applied to the MPC (Rawlings and Mayne 2009). Outside the terminal set, an N-step admissible set $\mathcal{X}_N$ can be found, for which the MPC is enforced to reach $\mathbb{X}_f$ in N steps if a terminal set constraint $\mathbf{x}_s(k+N) \in \mathbb{X}_f$ is added to $\mathscr{P}_s(\mathbf{x}_s(k))$ (Konig et al. 2014). Following Rawlings and Mayne (2009), the terminal set constraint is usually too complex for real-time implementations and can be omitted if the initial state lies inside a region of attraction $\bar{\mathcal{X}}_N$. Depending on the choice of the controller parameters, the region of attraction is defined as a sufficiently small subset of the admissible set $\bar{\mathcal{X}}_N \subseteq \mathcal{X}_N$.

A terminal set $\mathbb{X}_f$ for the given controller can be calculated following the algorithm proposed by Gilbert and Tan (1991), while Keerthi and Gilbert (1987) proposed an algorithm to calculate $\mathcal{X}_N$. Since the state vector only consists of three variables, $\mathbb{X}_f$ and $\mathcal{X}_N$ can be determined straightforward and depicted in a 3-D figure. Figure 4.4 depicts the terminal set $\mathbb{X}_f$ and the 25-step admissible set $\mathcal{X}_{25}$. Additionally, the state trajectories with initial states inside (blue) and outside (cyan) $\mathcal{X}_{25}$ are depicted as well.

As can be seen in the figure, both trajectories converge to the origin, while the blue trajectory reaches $\mathbb{X}_f$ in less than 25 samples. Since the initial state of the cyan trajectory is outside the admissible set, the controller needs 21 samples (equivalent to 210 ms) to reach $\mathcal{X}_{25}$. The slight increase of the rotational speed in both cases results from the input constraints which implies due to the compensation of the excessive torque.

Note that the proof of stability does not guarantee that engine stalling is avoided and that due to this reason, the load must lie within the feasible range of the powertrain (Unger et al. 2015). The stability of the control loop is also not effected by unpredicted disturbances, though the performance of the control loop is mainly depending on the controller tuning which may cause poor performance in case of prediction errors.

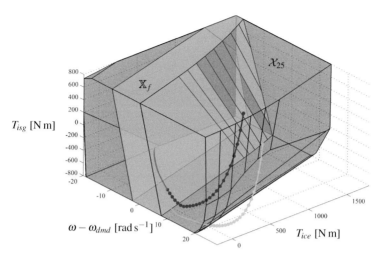

Fig. 4.4 Resulting terminal ($\mathbb{X}_f$) and 25-step admissible ($\mathcal{X}_{25}$) set obtained for the slave controller with operation point $\omega_{dmd} = 100$ rad/s

4.3.5 Master Controller

The goal of the master controller is to take SoC_{dmd} as a reference value and provide the demand set point values speed ω_{dmd} and ICE torque $T_{ice,dmd}$ for the slave controller. Due to the nonlinear behavior of the plant, optimal demand values are only obtained if the nonlinearities are considered (Lyshevski 2000). In the following, the concept and design of the master controller are developed in detail, and the convergence of the iterative approach is discussed.

4.3.5.1 Concept of Master Controller

A link between the applied load of the powertrain and the battery SoC must be established in the controller. Due to this reason, following the SoC model in Sect. 4.3.2, the nonlinear function k_I in Eq. (4.12) links the SoC to the ISG torque, which is proportional to the current of the battery. In each time and in each prediction step the model changes, which results in a nonlinear optimization problem to be solved in real time. A solution of the optimization problem can only be achieved by an iterative approach. The master controller is designed to consider a full load cycle duration in the optimization. In this book, a solution approach is proposed that consists of an inner iteration loop, a linear optimization $\mathscr{P}(\mathbf{x}_m)$ as well as an outer iteration loop. The flowchart in Fig. 4.5 depicts the concept schematically, where the variables indicate trajectories across the prediction horizon N_p. Note that the minimum prediction horizon is defined in such a way that the full cycle duration can be considered, while the real-time capability gives an upper limit for N_p. Due to consideration of the entire

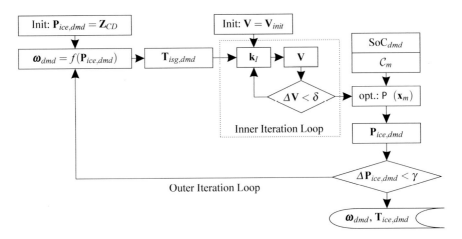

Fig. 4.5 Concept of the nonlinear master MPC

load cycle within the optimization, the energy storage can be fully exploited. In the following, the concept is discussed in more detail.

4.3.5.2 Inner Iteration Loop

A relaxation approach is used to consider the influence of the battery behavior. The implicit equations for the SoC model are evolved for optimization by using the predicted load trajectory $\mathbf{Z}_{CD} \in I\!\!R^{N_p}$ to initialize the ICE load trajectory $\mathbf{P}_{ice,dmd} \in I\!\!R^{N_p}$ and to obtain the optimal demand speed trajectory $\boldsymbol{\omega}_{dmd} \in I\!\!R^{N_p}$ from a characteristic map of the ICE. From Eq. (4.13) follows the ISG demand torque trajectory $\mathbf{T}_{isg,dmd} \in I\!\!R^{N_p}$, from which the initial $\mathbf{k}_I \in I\!\!R^{N_p}$ can be obtained using an initial voltage trajectory $\mathbf{V}_{init} \in I\!\!R^{N_p}$ as the actual voltage. Due to the ISG torque trajectory $\mathbf{T}_{isg,dmd} \in I\!\!R^{N_p}$, the current trajectory $\mathbf{I} \in I\!\!R^{N_p}$ by Eq. (4.12) affects the battery voltage $\mathbf{V}$ whose behavior is updated with the implemented battery model Eq. (4.17). Based on the determined $\mathbf{V}$, a more accurate voltage trajectory $\mathbf{V}$ is obtained by updating $\mathbf{k}_I$. This iteration is done until the voltage change $\Delta\mathbf{V}$ has converged to a small threshold value δ. A small threshold implies that the nonlinear plant behavior is considered in the SoC model.

4.3.5.3 MPC Formulation

The master controller, similar to the slave controller, is based on the augmentation of the obtained SoC model Eqs. (4.21) and (4.22) which follows to the incremental formulation (Wang 2009)

$$\mathbf{x}_m(k+1) = \mathbf{A}_m\mathbf{x}_m(k) + \mathbf{B}_m(k)\Delta u_m(k) + \mathbf{E}_m(k)\Delta z_m(k), \qquad (4.48)$$

$$\mathrm{SoC}(k) = \mathbf{C}_m\mathbf{x}_m(k), \qquad (4.49)$$

where $\mathbf{x}_m(k) = \begin{bmatrix} \Delta\mathrm{SoC}(k) & \mathrm{SoC}(k) \end{bmatrix}^T$ is the augmented state vector, Δu_m is the incremental input and Δz_m is the incremental disturbance. For the optimization, the objective function J_m to be optimized can be formulated by

$$J_m = Q_m \sum_{i=1}^{N_p-1} (\mathrm{SoC}_{dmd} - \mathrm{SoC}(k+i))^2 + R_m \sum_{i=0}^{N_c-1} (\Delta u_m(k+i))^2$$
$$+ \mathbf{x}_m(k+N_p)^T \mathbf{P}_m\mathbf{x}_m(k+N_p) \qquad (4.50)$$

where Q_m, R_m are weights. The weight $\mathbf{P}_m$ is calculated by the discrete algebraic Riccati equation

$$\mathbf{P}_m = \mathbf{A}_m^T\mathbf{P}_m\mathbf{A}_m - \mathbf{K}_m^T\left(\mathbf{R}_m + \mathbf{B}_m^T\mathbf{P}_m\mathbf{B}_m\right)\mathbf{K}_m + \mathbf{C}_m^T Q_m\mathbf{C}_m, \qquad (4.51)$$

where $\mathbf{K}_m = \left(\mathbf{R}_m + \mathbf{B}_m^T\mathbf{P}_m\mathbf{B}_m\right)^{-1}\mathbf{B}_m^T\mathbf{P}_m\mathbf{A}_m$. The stacked output prediction $\hat{\mathbf{Y}}_m$ follows then to

$$\hat{\mathbf{Y}}_m = \begin{bmatrix} \mathrm{SoC}(k+1) & \cdots & \mathrm{SoC}(k+N_p) & \mathbf{x}_m(k+N_p)^T \end{bmatrix}^T$$
$$= \mathbf{F}_m\mathbf{x}_m(k) + \boldsymbol{\Phi}_{m,u}\Delta\mathbf{U}_m(k) + \boldsymbol{\Phi}_{m,z}\Delta\mathbf{Z}_{CD}(k), \qquad (4.52)$$

with

$$\Delta\mathbf{U}_m(k) = \begin{bmatrix} \Delta u_m(k+1) & \ldots & \Delta u_m(k+N_c) \end{bmatrix}^T, \qquad (4.53)$$

$$\mathbf{F}_m = \left[(\mathbf{C}_m\mathbf{A}_m)^T \; (\mathbf{C}_m\mathbf{A}_m^2)^T \cdots \left(\mathbf{C}_m\mathbf{A}_m^{N_p-1}\right)^T \left(\mathbf{A}_m^{N_p}\right)^T \right]^T, \qquad (4.54)$$

$$\boldsymbol{\Phi}_{m,u} = \begin{bmatrix} \mathbf{C}_m\mathbf{B}_m(k|1) & \mathbf{0} & \ldots & \mathbf{0} \\ \mathbf{C}_m\sum_{i=1}^{2}\mathbf{B}_m(k|i) & \mathbf{C}_m\mathbf{B}_m(k|1) & \ldots & \mathbf{0} \\ \vdots & \vdots & \ddots & \vdots \\ \mathbf{C}_m\sum_{i=1}^{N_p-1}\mathbf{B}_m(k|i) & \mathbf{C}_m\sum_{i=1}^{N_p-2}\mathbf{B}_m(k|i) & \ldots & \mathbf{0} \\ \sum_{i=1}^{N_p}\mathbf{B}_m(k|i) & \sum_{i=1}^{N_p-1}\mathbf{B}_m(k|i) & \ldots & \sum_{i=1}^{N_p-N_c}\mathbf{B}_m(k|i) \end{bmatrix}, \qquad (4.55)$$

$$\boldsymbol{\Phi}_{m,z} = \begin{bmatrix} \mathbf{C}_m \mathbf{E}_m(k|1) & \mathbf{0} & \cdots & \mathbf{0} \\ \mathbf{C}_m \sum_{i=1}^{2} \mathbf{E}_m(k|i) & \mathbf{C}_m \mathbf{E}_m(k|1) & \cdots & \mathbf{0} \\ \vdots & \vdots & \ddots & \vdots \\ \mathbf{C}_m \sum_{i=1}^{N_p-1} \mathbf{E}_m(k|i) & \mathbf{C}_m \sum_{i=1}^{N_p-2} \mathbf{E}_m(k|i) \cdots & \mathbf{0} \\ \sum_{i=1}^{N_p} \mathbf{E}_m(k|i) & \sum_{i=1}^{N_p-1} \mathbf{E}_m(k|i) & \cdots \mathbf{E}_m(k|1) \end{bmatrix}, \qquad (4.56)$$

where $(k|i)$ denotes the time step k at prediction step i, $\Delta \mathbf{Z}_{CD}$ is the incremental disturbance trajectory obtained by the cycle detection and $\mathbf{F}_m$, $\boldsymbol{\Phi}_{m,u}$ and $\boldsymbol{\Phi}_{m,z}$ predicting the states. Due to the time varying process model Eqs. (4.21) and (4.22), the terms in $\boldsymbol{\Phi}_{m,u}$ and $\boldsymbol{\Phi}_{m,z}$ appear to be sum instead of a multiplication of $\mathbf{A}_m$.

In order to provide sufficient dynamics within the master controller, a sampling time of at least $t_{s,m} = 0.25$ s is beneficial, while the used computational capability limits the prediction horizon to $N_p = 100$. Due to this reason, cycle durations up to 250 s can only be considered by implementing a not equidistant prediction model. Following Eq. (4.18), which contains the sampling time $t_{s,m} = 0.25$ s of the master controller, $t_{s,m}$ is adapted for the prediction instants. Note that this provides the possibility to achieve dynamic control with a sampling time of $t_{s,m} = 0.25$ s, while a wide prediction horizon is covered. The inaccuracy due to the larger sampling intervals toward the end of the prediction horizon may have negligible influence, since the mean discharge—mainly important—is considered. A multiplication factor for each single SS-model can be used to implement the different sampling times of the prediction within the cost function J_m (Unger et al. 2015).

Under consideration of the set of constraints Eq. (4.34), the formal optimization problem can be denoted by

$$\mathscr{P}(\mathbf{x}_m(k)) : \quad \Delta \mathbf{U}_m^*(\mathbf{x}_m(k)) = \arg \min J_m(\mathbf{x}_m(k), \Delta \mathbf{U}_m),$$
$$\text{s.t. } \mathcal{C}_m : \mathbf{M}_m \Delta \mathbf{U}_m \le \boldsymbol{\gamma}_m$$
$$\text{with} \quad \mathbf{M}_m = \begin{bmatrix} \mathbf{M}_{m,\Delta u} \\ \mathbf{M}_{m,u} \\ \mathbf{M}_{m,y} \end{bmatrix}, \quad \boldsymbol{\gamma}_m = \begin{bmatrix} \boldsymbol{\gamma}_{m,\Delta u} \\ \boldsymbol{\gamma}_{m,u} \\ \boldsymbol{\gamma}_{m,y} \end{bmatrix} \qquad (4.57)$$

referring to rate, input and output constraints, respectively. Similar to the slave controller, the real-time optimization is realized using the FMPC algorithm proposed by Wang and Boyd (2010).

4.3.5.4 Outer Iteration Loop

As a result, the optimization problem Eq. (4.57) provides the incremental trajectory of the ICE torque $\Delta \mathbf{U}_m^*$ that minimizes the SoC deviation from SoC_{dmd} at a defined

ω_{dmd}. Calculating the demand ICE torque by

$$T_{ice,dmd}(k+i) = T_{ice}(k) + \sum_{l=1}^{i} \Delta u_m^*(l) \tag{4.58}$$

and taking ICE load constraints into consideration, the demand ICE load $\mathbf{P}_{ice,dmd}$ can be obtained by

$$P_{ice,dmd}(k) = \omega_{dmd}(k)T_{ice,dmd}(k),$$
$$\text{s.t. } P_{ice,min} \leq P_{ice,dmd}(k) \leq P_{ice,max}. \tag{4.59}$$

The outer iteration loop is obtained by solving the inner iteration loop and the optimization Eq. (4.57) anew with the obtained ICE load $\mathbf{P}_{ice,dmd}$ until $\Delta P_{ice,dmd}$ is converged to a defined threshold γ or the maximal computation time is reached. Note that the ICE load represents another nonlinearity of the master controller that is considered due to iteration loop.

The master controller equals a power controller that is capable of controlling the SoC, speed as well as ICE and ISG torque, if none of the powertrain limits are violated. Nevertheless, the optimal trajectories can only be achieved if the system is controllable, which is assumed for any load request.

4.3.5.5 Convergence of the Concept

The Banach fixed-point theorem shows convergence for iteration loops, if the Lipschitz condition is fulfilled and the sequence is Cauchy. Using Eq. (4.12) and (4.17), the inner iteration loop can be denoted by an implicit equation $\mathbf{V} = f_v(\mathbf{T}_{isg,dmd}, \boldsymbol{\omega}_{dmd}, \text{SoC}, \vartheta_{batt}, \mathbf{V}) = f_v(\zeta, \mathbf{V})$. Similarly, with Eqs. (4.11)–(4.13), (4.18) and (4.19) as well as the MPC, the outer iteration loop is expressed by a function $\mathbf{P}_{ice,dmd} = f_p(\mathbf{P}_{load}, \text{SoC}, \vartheta_{batt}, \mathbf{P}_{ice,dmd}) = f_p(\xi, \mathbf{P}_{ice,dmd})$. Both equations are assumed to be feasible in the operation range, since the output voltage $\mathbf{V}$ as well as demand ICE load $\mathbf{P}_{ice,dmd}$ are bounded by constraints.

The proof for convergence can be done by standard linear MPC literature (c.f. e.g. Gunderson 2010, Istratescu 2001, Vidyasagar 2002) and is similar for both iterations as given in the following.

Assumption 1 Both functions f_v and f_p satisfy the Lipschitz condition such that

$$|f_v(\zeta, \mathbf{V} + \Delta\mathbf{V}) - f_v(\zeta, \mathbf{V})| \leq L_v |\Delta\mathbf{V}|, \tag{4.60}$$

$$\left|f_p(\xi, \mathbf{P}_{ice,dmd} + \Delta\mathbf{P}_{ice,dmd}) - f_p(\xi, \mathbf{P}_{ice,dmd})\right| \leq L_p \left|\Delta\mathbf{P}_{ice,dmd}\right|, \tag{4.61}$$

with $L_v, L_p > 0$ hold. Therefore, if f_v and f_p are continuous and satisfy Eqs. (4.60) and (4.61), respectively, it follows that each of f_v and f_p have unique solutions (Gao 2012; Unger et al. 2015).

Denoting $\mathbf{X}$ instead of $\mathbf{V}$ and $\mathbf{P}_{ice,dmd}$, respectively, with Assumption 1 and mathematical induction, Lemma 1 follows:

Lemma 1 *For all $k \in \mathbb{N}_{>0}$, arbitrary $\mathbf{X}_0$ and Lipschitz constant $L \in [0, 1)$, $\|\mathbf{X}_{k+1} - \mathbf{X}_k\|_2 \le L^k \|\mathbf{X}_1 - \mathbf{X}_0\|_2$ holds.*

Proof Proceeding using mathematical induction, the base case holds:

$$\|\mathbf{X}_2 - \mathbf{X}_1\|_2 = \|f(\mathbf{X}_1) - f(\mathbf{X}_0)\|_2 \le L \|\mathbf{X}_1 - \mathbf{X}_0\|_2 \tag{4.62}$$

Then, supposing the statement holds for some $k \in \mathbb{N}_{>0}$, the induction hypothesis follows to

$$\|f(\mathbf{X}_{k+1}) - f(\mathbf{X}_k)\|_2 \le L \|\mathbf{X}_{k+1} - \mathbf{X}_k\|_2 \tag{4.63}$$
$$\le L L^k \|\mathbf{X}_1 - \mathbf{X}_0\|_2 \tag{4.64}$$
$$= L^{k+1} \|\mathbf{X}_1 - \mathbf{X}_0\|_2 , \tag{4.65}$$

which proves the Lemma by the principle of mathematical induction (Gunderson 2010). □

Based on Lemma 1, the sequence can be shown to be Cauchy.

Lemma 2 *Let $M \in \mathbb{R}^{N_p}$ be a metric space. The sequence $\{\mathbf{X}_k\}$ in M is a Cauchy sequence and therefore converges with a Lipschitz constant $L \in [0, 1)$ to a limit $\mathbf{X}^*$ in M (Istratescu 2001).*

Proof Let $m, n \in \mathbb{N}_{>0}$ such that $m > n$. Using the Triangle Inequality, Lemma 1 and the Geometric Series, the following can be denoted:

$$\|\mathbf{X}_m - \mathbf{X}_n\|_2 \le \|\mathbf{X}_m - \mathbf{X}_{m-1}\|_2 + \cdots + \|\mathbf{X}_{n+1} - \mathbf{X}_n\|_2 \tag{4.66}$$
$$\le L^n \|\mathbf{X}_1 - \mathbf{X}_0\|_2 \sum_{k=0}^{m-n-1} L^k \tag{4.67}$$
$$\le L^n \|\mathbf{X}_1 - \mathbf{X}_0\|_2 \sum_{k=0}^{\infty} L^k \tag{4.68}$$
$$= L^n \|\mathbf{X}_1 - \mathbf{X}_0\|_2 \left(\frac{1}{1-L}\right). \tag{4.69}$$

Let $\varepsilon > 0$ arbitrary, a large $N \in \mathbb{N}_{>0}$ can be found such that

$$L^N < \frac{\varepsilon (1-L)}{\|\mathbf{X}_1 - \mathbf{X}_0\|_2} \tag{4.70}$$

is satisfied, and

$$\|\mathbf{X}_m - \mathbf{X}_n\|_2 \le L^n \|\mathbf{X}_1 - \mathbf{X}_0\|_2 \left(\frac{1}{1-L}\right) < \varepsilon \tag{4.71}$$

follows for m, n large enough. Since $\varepsilon > 0$ is arbitrary, the sequence is proven to be Cauchy. □

Using Assumption 1, Lemmas 1 and 2, Theorem 1 follows:

Theorem 1 (Banach Fixed-Point Theorem (Banach 1922)) *A map $f : M \rightarrow M$ is called contraction mapping on M if there exists a Lipschitz constant $L \in [0, 1)$ such that*

$$\| f (\mathbf{X}_{k+1}) - f (\mathbf{X}_k) \|_2 \leq L \| \mathbf{X}_{k+1} - \mathbf{X}_k \|_2, \ \forall k \in \mathbb{N}_{>0} \qquad (4.72)$$

with $\mathbf{X}_k = f (\mathbf{X}_{k-1})$. If

$$L^N < \frac{\varepsilon (1 - L)}{\| \mathbf{X}_1 - \mathbf{X}_0 \|_2}, \quad \varepsilon > 0, \qquad (4.73)$$

is satisfied for a large $N \in \mathbb{N}_{>0}$, the following inequalities hold:

$$\frac{\| \mathbf{X}_{k+1} - \mathbf{X}_k \|_2}{\| \mathbf{X}_1 - \mathbf{X}_0 \|_2} \leq L^k < \frac{\| \mathbf{X}_2 - \mathbf{X}_1 \|_2}{\| \mathbf{X}_1 - \mathbf{X}_0 \|_2} < 1. \qquad (4.74)$$

Theorem 1 is fulfilled if $\exists L \in [0, 1)$.

Convergence of iteration loops. Based on Theorem 1, $0 \leq L < 1$ can be shown for both iteration loops separately by evaluating the closed set of possible configurations. Both iterations converged to the true value $\mathbf{V}^*$ (see Fig. 4.6) and $\mathbf{P}^*_{ice,dmd}$ (see Fig. 4.7), respectively.

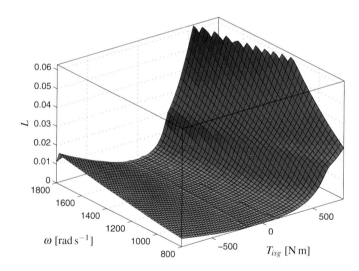

Fig. 4.6 Inner iteration loop: resulting contraction map shows that the Lipschitz constant L exists and is $0 \leq L < 1$ for the entire operation range

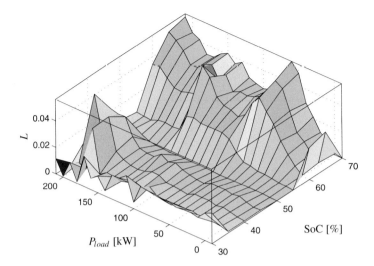

Fig. 4.7 Outer iteration loop: resulting contraction map shows that the Lipschitz constant L exists and is $0 \leq L < 1$ for the entire operation range

4.4 Load and Cycle Prediction for Non-Road Machinery

The driver exclusively influences the load trajectory and driving patterns of non-road machinery (Frank et al. 2012). Exact prediction of future load demands is in general very difficult or sometimes impossible. Nevertheless, in the following, two statistical approaches are introduced, which are able to predict the future load trajectories with sufficient accuracy for the usage in non-road vehicles. A short-term load trajectory $\mathbf{Z}_{LP}$ for use in the slave MPC and a long term cycle prediction $\mathbf{Z}_{CD}$ for use in the master MPC are proposed in detail. The cycle prediction is based on a cycle detection that is similar to the approach proposed by Mayr et al. (2011a).

4.4.1 Short-Term Load Prediction

The idea of short-term load prediction is to detect critical load demands of, e.g., wheel loaders at digging or at reversion, in order to permit a control action in advance (e.g., increasing the rotational speed in advance to a load peak acting on the powertrain). For example, at reversion of the vehicle, if the load requirement is especially high shortly after direction change and vehicle acceleration, the load prediction offers the possibility to avoid engine stalling or a speed undershoot. Driveability and handling capacity may be increased, which is especially of interest for the industry. In the following, the theory of Bayesian inference is shortly reviewed before the methodology is discussed in detail. Based on the wheel loader, the accuracy of the methodology is validated based on real data measured on a real wheel loader.

4.4.1.1 Bayesian Inference

In the field of mathematical statistics, the Bayesian inference (BI) is an important technique to update a probability of a hypothesis (H) as new evidence (E) is available. Especially for dynamic data sequence analysis, the BI has a decisive role. Based on Bayes' rule, the posterior probability is calculated using the a-priori probability and the likelihood function, respectively. The latter two can be obtained from a probability model describing the data to be observed. The posterior probability is calculated using Bayes' theorem Eq. (4.75)

$$P(H \mid E) = \frac{P(E \mid H) \cdot P(H)}{P(E)}, \tag{4.75}$$

where | denotes a conditional probability, $P(H)$ and $P(H \mid E)$ are prior and posterior probability, respectively, $P(E)$ is the likelihood function and $P(E \mid H)$ is the conditional probability of observing evidence E if hypothesis H is given. Note here that the posterior probability is proportional to $P(H)$ and $P(E \mid H)$, since these values appear in the numerator of Eq. (4.75). Using the theory of Bayesian inference, the load prediction methodology can be developed.

4.4.1.2 Methodology of Load Prediction

The actual vehicle inputs u_i, such as accelerator angle α, driving speed n_{GW} and load P_{load} offer the only information about the vehicle state that can be used to predict the future load demand. In case of a specific input state configuration, a probability can be calculated that provides statistical evidence about the future load demand depending on the actual vehicle state. For that purpose, inputs and outputs are assigned to discrete classes ϕ_{ij} in the possible input/output range in order to be able to apply the theory of the discrete Bayesian inference. Based on Bayes' rule, the a posteriori probability for a certain load class ψ_{1j} given a certain input configuration $\underline{\Phi}(k)$ can be calculated. A-priori probability and the likelihood function are required (Marques De Sá 2003) and can be obtained from training data (Box and Tiao 2011). Note that the accuracy is directly dependent on the training data, and therefore all relevant information needs to be included in the training data. Determining the a-priori probability and the likelihood function based on training data shifted backwards in time, the basis to calculate the a posteriori probabilities $P\left(\psi_{hj}|\underline{\Phi}\right)$ for further prediction steps h is given. The flowchart of the methodology is depicted in Fig. 4.8, where the rule of Bayes to calculate $P\left(\psi_{hj}|\underline{\Phi}\right)$ is denoted.

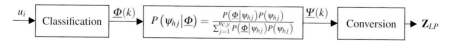

Fig. 4.8 Flowchart of the load prediction process

4.4.1.3   Prediction of the Future Load Demand

Since a load trajectory is required by the controllers, the probabilities $P\left(\psi_{hj}|\Phi\right)$ need to be converted into a load value. A simple approach to extract the load value is to choose the highest load class probability $\max\left(P\left(\psi_{hj}|\Phi\right)\right)$ and use the corresponding maximum load value associated by the class for each prediction step h. Note that in case of large load classes, the resolution is rough and the prediction may have insufficient accuracy. This can be avoided by choosing enough output classes or an average class load value. The short-term load trajectory $\mathbf{Z}_{LP}$ follows therefore by storing the corresponding load value $z_{LP,h}$ of the class in a vector

$$\mathbf{Z}_{LP} = \left[z_{LP,1} \dots z_{LP,n_p}\right]^T. \tag{4.76}$$

The incremental notation required by the MPC follows by

$$\Delta \mathbf{Z}_{LP} = \left[\Delta z_{LP}(1) \dots \Delta z_{LP}(n_p)\right]^T,$$

$$\text{with} \quad \Delta z_{LP}(i) = \begin{cases} z_{LP,1} - P_{load}(k) & \text{if } i = 1 \\ z_{LP,i} - z_{LP,i-1} & \text{else} \end{cases}. \tag{4.77}$$

4.4.1.4   Validation of the Load Prediction

For validation of the load prediction, training data were recorded on a corresponding wheel loader at operation for an eight hour duration. An analysis showed the following input configuration $\underline{u}$ to be significant for the load prediction

$$\underline{u} = \left[\alpha \quad |n_{GW}| \quad P_{load} \quad \Delta^2\left(\text{filt}\left(P_{load}\right)\right)\right], \tag{4.78}$$

where Δ^2 denotes the dual backwards difference operator and filt() is a low pass filter. The selection includes the driver information (α), an indication for the overall power consumption ($|n_{GW}|$) and the actual state of the vehicle (P_{load}, $\Delta^2\left(\text{filt}\left(P_{load}\right)\right)$). Note that the filter input corresponds to the mechanical inertia and needs to be tuned in such a way that strong gradients are not too much delayed, but that noise and small load peaks are avoided. A selection of 6 input classes $n_{c,i}$ for α, 4 for $|n_{GW}|$, 24 for P_{load} and 6 for the filter input showed to be adequate, while 24 output classes $n_{c,y}$ have been chosen to achieve sufficient resolution for the load prediction.

Figure 4.9 shows the scaled input signals (subplot two) corresponding to the measured load signal (subplot one, red) and the predicted short-term load trajectories across a horizon of $n_p = 25$ samples (subplot one, blue). The load sequence is previously unknown by the load prediction, but though, accurate predictions are achieved at increasing load gradients (c.f. around second 2 and 9). This is plausible since accelerator position, filter input and driving speed indicate a clear load demand. The largest part of the overall load is in general consumed by the drivetrain. Around

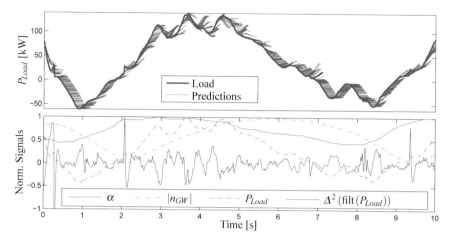

Fig. 4.9 Validation of the load prediction based on a real load sequence

second 6, the driving speed is increasing although the accelerator position is around
60 % and decreasing. At this point, a serious indication is impossible, since not even
the filter input can provide an additional indication, and therefore, the actual load is
in principle predicted.

Between second 7 and 8, negative loads appear, which are not critical to engine
stalling. A significant prediction mismatch results from a low resolution obtained
by the simplification of assigning only one load class for the entire negative load
range. Note that an increase in the number of output load classes would enhance the
prediction accuracy in the negative load range.

4.4.2   Cycle Detection

Dockside cranes or wheel loaders are mostly used in cyclical operations, which yield
recurrent load demands that can be used to predict the future cycles. The goal is to
predict the load trajectory by detecting a recurrent load cycle $\mathbf{z}_{cyc} \in \mathbb{R}^{s_{cyc}}$ within the
past load signal and to provide a disturbance trajectory $\mathbf{Z}_{CD} = \left[\mathbf{z}_{cyc}^T \ldots \mathbf{z}_{cyc}^T \right]^T \in$
$\mathbb{R}^{N_p}$ with sampling time $t_{s,m}$ for the master controller based on the obtained cycle
information. Correlation analysis provides the necessary theory to find the recurrent
cycles within the past load signal. In the following, the theory of correlation analysis
is therefore reviewed before the methodology of the cycle detection is presented in
detail. Based on a real load signal obtained from a wheel loader, the methodology is
validated to show the possibilities.

4.4.2.1 Cross Correlation Function

Commonly, in signal processing, tasks consist of finding short signals within a long
signal. A measure for the similarity of two signals is given by the cross correlation,
which has applications over all natural sciences. It is used in, e.g., pattern recognition, single particle analysis, electron tomography, averaging, cryptanalysis, or
neurophysiology. In terms of statistical analysis, the term cross correlation refers to
the correlation between the values of two random vectors $\mathbf{x}$ and $\mathbf{y}$, while a special
case is the correlation of the values of one vector $\mathbf{x}$ with itself, which is referred to
as auto correlation (ACF). The cross correlation function is defined by

$$\hat{R}_{xy}(l) = \sigma \sum_{k=1}^{N} x(k)\, y(k - l),\tag{4.79}$$

where x, y are the entries of the vectors $\mathbf{x}$ and $\mathbf{y}$ respectively, N is the signal length
and σ is a scaling factor. Note here that in the field of statistical analysis, the scaling
factor σ is usually included in the definition in order to obtain correlations between
-1 and $+1$. Based on the cross correlation function, the methodology of the cycle
detection can be developed in the following.

4.4.2.2 Identification Methodology for Recurrent Load Cycles

Mayr et al. (2011a) proposed an algorithm to detect the cycle duration $t_{cyc} = s_{cyc}t_{s,m}$
of a recurrent load cycle within the past load signal $\mathbf{y}_{CD} \in \mathbb{R}^{\geq 2 \cdot s_{cyc}}$ based on the
auto correlation function. The first local maximum in the ACF is used to identify
the cycle. One disadvantage of the ACF is that in case of a changing load cycle, a
detection of the cycle with an acceptable time delay is difficult. In case of cycles
with significantly different cycle times, the shape of the first local maximum is not
formed clearly, which makes a secure detection impossible.

In this book, the cross correlation function $\hat{R}_{xy,i}$ between $\mathbf{y}_{CD}$ and different parts
$\mathbf{x}_{CD,i}$, $i = 1, 2, 3, \ldots$, which are systematically taken from $\mathbf{y}_{CD}$, is used to adapt
the ACF approach in order to avoid this disadvantage. The size of the parts $\mathbf{x}_{CD,i}$
are within the scope of the possible cycle durations within $\mathbf{y}_{CD}$, due to which reason
a cycle change can be detected sufficiently fast. A cycle is detected and included
in $\mathbf{y}_{CD}$, if more than two local maxima of $\hat{R}_{xy,i}$ lie within a small range around
$\hat{R}_{xy,i} = 1$, and the distance between the local maxima defines the corresponding
cycle time (Unger et al. 2015).

Once a cycle is detected, the past s_{cyc} load values form the load cycle $\mathbf{z}_{cyc}$, which
is used to build the incremental notation of the future load trajectory $\mathbf{Z}_{CD}$ used in
the master controller by

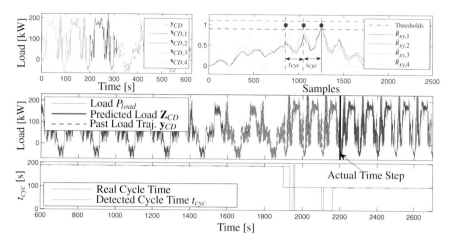

Fig. 4.10 Cycle detection methodology shown on a sequence of two different load cycles

$$\Delta \mathbf{Z}_{CD}(k) = \left[\Delta z_{CD}(1) \cdots \Delta z_{CD}(N_p) \right]^T,$$

$$\text{with} \quad \Delta z_{CD}(i) = \begin{cases} Z_{CD}(i) - P_{load}(k) & \text{if } i = 1 \\ Z_{CD}(i) - Z_{CD}(i-1) & \text{else} \end{cases}. \quad (4.80)$$

4.4.2.3 Validation of the Methodology

A more sophisticated representation of the methodology is depicted in Fig. 4.10, where the cycle detection is demonstrated on real load measurement from a wheel loader. The measured signal (blue), the past load trajectory $\mathbf{y}_{CD}$ (orange) and the predicted disturbance trajectory $\mathbf{Z}_{CD}$ (red) are drawn in the middle subplot, while the true (blue) with the detected (green) cycle time t_{cyc} are given in the subplot below. It can be seen that almost 3 cycles are required to detect the cycle change, which is sufficiently fast, since a confident prediction needs at least 3 occurred cycles. Note that, though, the new cycle is only a short part of $\mathbf{y}_{CD}$, the cycle time can be detected correctly.

In the first row of subplots, the past load signal $\mathbf{y}_{CD}$ (orange) and the different parts $\mathbf{x}_{CD,i}$ (blue, magenta, green, cyan) are depicted on the left side, while the corresponding CCFs $\hat{R}_{xy,i}$ (blue, magenta, green, cyan) between $\mathbf{x}_{CD,i}$ and $\mathbf{y}_{CD}$ are shown on the right side. The blue points indicate that the local maxima lie inside the thresholds (black dashed lines) and match for the different $\hat{R}_{xy,i}$. Consequentially, a cycle is detected.

Chapter 5
Application Example: Wheel Loader

Abstract The Chaps. 2, 3, and 4 discussed the methodologies to estimate the battery SoC with high accuracy as well as to control a parallel hybrid electric powertrain in non-road vehicles. In this chapter, the results obtained by these methodologies should be presented. First, the hardware of the used real powertrain and test bed including a battery simulator, respectively, are described. Second, simulation and real measurement results are presented in terms of dynamics, overall optimality, and efficiency improvement due to the proposed EMS.

Keywords Non-road hybrid electric vehicle (HEV) · Nonlinear model predictive control · Load and cycle prediction · Stability · Convergence · Real-time control

5.1 Hardware Configuration of the Hybrid Powertrain Test bed

Non-road vehicles are usually equipped with more powerful engines than they are registered with, which in this context means that they are restricted in power. At the test bed, a real diesel electrical parallel hybrid powertrain comprising a 290 kW ICE (limited at $P_{ice,max} = 215$ kW) and a 120kW ISG is set up. A torque sensing shaft connects the test bed dyno with the powertrain in order to apply the load. Instead of a real battery, a battery simulator supplies the required power for the ISG where an accurate LMN battery model of the lithium-iron-phosphate chemistry-based battery module with 630 V nominal voltage and $\pm$ 200 A maximal current (192S2P configuration) is implemented (see Sect. 3.6). Note that the powertrain is also modeled and used for real-time simulation.

A dSpace DS1006 platform is used to control the test bed measurements as well as real-time simulations. The platform runs the standard fixed step-size solver for which the controller is compiled, while the optimization problem is solved using the primal barrier interior-point method in theFMPC algorithm proposed by Wang and

© The Author(s) 2016
J. Unger et al., *Energy Efficient Non-Road Hybrid Electric Vehicles*,
SpringerBriefs in Applied Sciences and Technology,
DOI 10.1007/978-3-319-29796-5_5

Fig. 5.1 Parallel hybrid electric powertrain at the test bed

Boyd (2010). Further details on algorithm and computation times can be found in
Wang and Boyd (2010) and in Richter et al. (2010), Zeilinger et al. (2011). The proposed controller can be parametrized by the number of iterations in the FMPC, which
are properly chosen to $n_{iter,FMPC,slave} = 10$ and $n_{iter,FMPC,master} = 3$, respectively,
the number of inner ($n_{iter,inner} = 10$) and outer ($n_{iter,outer} = 3$) iteration loops of
the master controller and the prediction horizons for both MPCs in order to achieve
lower computation times. Based on the given selections, the runtime limit of the
platform is fully exploited, but is not exceeded. Relevant information between platform and supervisory test bed control system is exchanged with a CAN interface,
while a sampling time of $t_{s,s} = 10$ ms is used for the interface as well as for all
measurements. Figure 5.1 shows a picture of the test bed application.

5.2   Energy Management in Wheel Loaders

The feasibility of the control concept is validated by carrying out real test bed measurements using the real powertrain described previously. Important to show is that
the controller is able to cope with the high dynamic load transients appearing in
non-road applications, while emissions and fuel consumption are reduced simultaneously. A critical point of the control is, if constraints are active. For the purpose
of reasonable results, the controller is tuned based on simulations, before test bed
measurements are carried out. In the following, the controller penalties are discussed
before the simulations are presented and measurement results are discussed.

5.2.1 User-Defined Tuning of the Controller Penalties

The selection of the penalties used in the optimization problems $\mathscr{P}_s(\mathbf{x}_s(k))$ and $\mathscr{P}_m(\mathbf{x}_m(k))$ is referred to as tuning and influences the behavior of the controller. In order to achieve maximal efficiency of the powertrain, the electrical energy conversion (EEC) must be kept at a minimum, since efficiency and EEC are directly linked by the conservation of energy principle. To this end, the penalties need to be set in such a way that dynamic requirements of the powertrain are provided in any case and the EEC is minimized simultaneously.

The SoC deviation from the demand SoC has direct influence on the EEC and can be penalized with the state penalty Q_m. By choosing $Q_m = 0.1$ (*more EEC*) and $Q_m = 10$ (*less EEC*), the influence of the SoC deviation can be seen and compared. In terms of dynamic and a good speed controlling performance, the state penalty Q_s plays a major role and is therefore chosen significantly higher than the control $\mathbf{R}_s$ and input $\tilde{\mathbf{R}}_s$ penalties. Desirable is also a smooth dynamic behavior of the ICE, since strong torque gradients cause higher emissions. This is achieved by penalizing the ICE torque increments more than the ISG torque increments. Note that this is inconsistent with the required dynamics, but due to the optimization, solved in an optimal manner. In contrast to the slave dynamics, the master controller has inferior influence on the fast control dynamics, due to which the control penalty R_m is chosen appropriately small. The parameters of the model and controller, respectively, which are used for simulations and test bed measurements, are summarized in Table 5.1 to give a better overview.

5.2.2 Simulation Results

The simulations are executed on the real-time platform, where a real-time model of the powertrain is implemented. Main interest of the simulations is to validate the controller tuning in terms of the optimality of the controller set points, while a ICE torque gradient restriction of $\Delta T_{ice,max} = 1000$ Nm/s is applied to limit the ICE dynamics.

Based on the dimensionless energetic efficiency map η_{ice}, in Fig. 5.2, the simulation results are compared with a real load cycle of a conventional operated powertrain (black $\times$). The results of the hybrid powertrain are inserted for the hybrid strategies with small (purple, $Q_m = 0.1$) and large (dark green, $Q_m = 10$) state penalty in order to see the difference between the strategies. As can be seen, the distribution of the operation points is limited to the optimal characteristic line (dark yellow) with the lowest specific fuel consumption for $Q_m = 0.1$, while for $Q_m = 10$ an increased distribution is obtained due to the increased usage of the ICE to compensate the transient load.

Table 5.1 Summary of the used controller parameters

Description	Parameter	Nominal value
Slave controller sampling time	$t_{s,s}$	0.01 s
Master controller sampling time	$t_{s,m}$	0.25 s
Master controller prediction horizon	$N_p = N_c$	100
Master controller state penalties	Q_m	0.1, 10
Master controller control penalty	R_m	0.01
Slave controller prediction horizon	$n_p = n_c$	25
Slave controller state penalty	Q_s	100
Slave controller control penalties	$\mathbf{R}_s$	diag[10 2]
Slave controller input penalties	$\bar{\mathbf{R}}_s$	diag[0.5 1]
Engine operation speed	$\omega_{min} \ldots \omega_{max}$	1000 ... 2000 rpm
Battery state of charge operation limits	$\mathrm{SoC}_{min} \ldots \mathrm{SoC}_{max}$	20 ... 80 %
Battery current limits	$I_{min} \ldots I_{max}$	$-200 \ldots 200$ A
Battery voltage limits	$V_{min} \ldots V_{max}$	550 ... 700 V
Maximal ICE power constraint	$P_{ice,max}$	215 kW
Battery capacity	$Q_{c,batt}$	8.8 A h
Total ICE moment of inertia	Θ	10 kg m^2
Time constant of ICE model	τ_{ice}	0.1 s
Time constant of ISG model	τ_{isg}	0.05 s

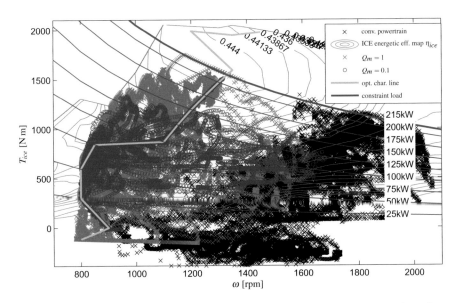

Fig. 5.2 Simulation results for two different hybrid strategies shown by the ICE operation points

Compared to the conventional powertrain, a significant decrease of the mean speed is observable due to the optimal operation set points, which proves that the down-speeding strategy is considered directly as proposed. The outliers at $\omega \approx 1500$rpm and $T_{ice} \approx 750$ N m are caused by the driver request $\omega_{min,driver}$.

5.2.3 Experimental Results

The experimental results focus mainly on the three following critical points and are presented in detail in the following:

1. Dynamic feasibility under consideration of all relevant constraints of the power-train and hybrid strategies such as downspeeding and phlegmatization.
2. Reduction of exhaust emissions and fuel consumption compared to the conventional powertrain due to optimal control.
3. The benefit of a full exploitation of the energy storage capabilities by use of the cycle detection.

5.2.3.1 Dynamic Feasibility

Phlegmatization incises the dynamics of the powertrain drastically, especially at an optimal (low) rotational speed, but provides a large potential to reduce exhaust emissions significantly (Nüesch et al. 2014). Feasibility mainly depends on the phlegmatization rate, though. On the test bed, different phlegmatization rates between $\Delta T_{ice,max}$ between 500 Nm/s and 5000 Nm/s are measured, while a rate of 500 Nm/s showed a significant emission reduction. Nevertheless, cycles with higher load gradients may not be feasible using 500 Nm/s, and therefore a time-variant phlegmatization rate is applied to the powertrain: In general, the ICE torque is limited by $\Delta T_{ice,max} = 500$ Nm/s, but after an active constraint for the past 15 samples, it is relaxed to $\Delta T_{ice,max} = 1500$ Nm/s. Note that an effective maximum of 1350 Nm/s is reachable, which is an appreciable limitation. Furthermore, the load prediction is used to avoid engine stalling on high load peaks and to provide sufficient engine dynamics. The influence of the load prediction on the reduced dynamics of the ICE is presented in Fig. 5.3, where the enabled load prediction (subscript LP) is compared to the disabled case. The first subplot of Fig. 5.3 shows the rotational speed trajectories of the actual speed ω and the speed set point ω_{set} for enabled (subscript LP) and disabled load prediction. Marked point (a) shows that in case of an enabled LP, ω_{set} is raised prior to the load peak. At marked point (b) engine stalling occurs for disabled LP due to insufficient ICE dynamics.

The time-variant phlegmatization rate is observable in the second subplot (see marked point (c)), where the trajectories of the ICE torques are depicted. Point (b) shows furthermore that $T_{ice,LP}$ is increased prior to a load peak, due to which engine stalling is avoided. The missing constraints for the driveability limit of the ICE

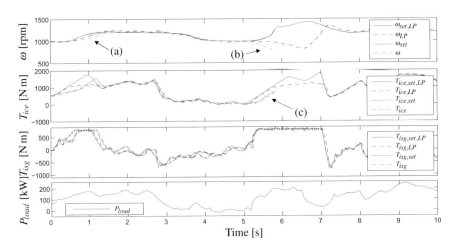

Fig. 5.3 Influence of an enabled (subscript LP) load prediction on the powertrain operation

are observable at higher torque levels, where a large mismatch occurs. Prior to the testruns, driveability limit information was not available and could not be considered. Nevertheless, the structured constraints as introduced in Sect. 4.3.3 provide to integrate any additional limits easily.

The ISG torque signals are shown in the third subplot, where it can be observed that in terms of dynamics, the load prediction smooths T_{isg} as well as reduces the dynamics of the signal in general. The last subplot shows the corresponding load trajectory for P_{load}. Note that in order to avoid engine stalling with disabled LP, the speed must be increased, which also increases fuel consumption due to friction losses. Hence, the load prediction enables to further reduce fuel consumption although the load demand is unknown and ICE dynamics are restricted.

5.2.3.2 Fuel Consumption and Exhaust Emissions

In order to achieve a suitable comparison between the hybrid and conventional powertrain, real load cycles are measured at the test bed with the same experimental setup. For that purpose, four representative real load cycles are extracted from a large data set of field measurements obtained from common wheel loaders representing common applications. In Table 5.2, more details about the cycles are summarized, which test bed measurements are used for comparison of fuel consumption and raw exhaust emissions.

The ECU of the used ICE provides two series controllers that take the speed or ICE torque demand as reference value. For the reference measurements on the conventional ICE, the speed controller is used to control the speed demand, which is proportional to the accelerator position. In contrast, the series torque controller is used for the hybrid measurements to execute the set point torque of the hybrid

Table 5.2 Basic data of the used load cycles

Cycle	Duration (s)	Mean load (kW)	Mean speed (rpm)
1	38.48	70.1	1236
2	54.15	61.3	1215
3	189.84	58.4	1360
4	90.44	66.5	1277

Table 5.3 Resulting reductions for fuel consumption and exhaust emissions obtained with enabled load and cycle prediction

Cycle	Strategy	Fuel (%)	CO (%)	HC (%)	NOx (%)	Soot (%)
1	$Q_m = 10$	−2.04	−50.16	−9.64	8.12	−21.00
1	$Q_m = 0.1^{(a)}$	−2.98	−54.98	−13.46	8.44	−31.01
2	$Q_m = 10$	−2.00	−55.09	−14.99	7.45	−27.25
2	$Q_m = 0.1^{(b)}$	−1.39	−56.37	13.52	8.99	−29.81
3	$Q_m = 10$	−8.54	−71.76	−11.23	7.10	−17.74
3	$Q_m = 0.1^{(b)}$	−9.14	−70.70	−3.65	7.64	−7.47
4	$Q_m = 10$	−3.79	−51.26	−11.59	9.27	−19.02
4	$Q_m = 0.1^{(a)}$	−4.26	−53.76	−11.76	9.52	−22.35

Relative change is calculated from the conventional powertrain. Small mean discharges (a) and charges (b) of the battery are not compensated in the results

controller on the shaft. Since the proposed control concept is supervisory in terms of the components, the parametrization of both series controllers is not changed in any way.

The compensation of any SoC change in the battery is an important topic, which plays a significant role, especially at short load cycles. Many works address the issue to compensate the deviations correctly, e.g., Sciarretta et al. (2004) and Johnson et al. (2000). In this book, in order to avoid the problem, the values for fuel consumption and raw exhaust emissions are obtained by repeating any load cycle for 10 times and determining the average value. Almost all cycles reached the initial SoC after the 10th cycle, due to what the SoC deviation can be neglected. Nevertheless, any significant deviations are marked in Table 5.3, where the results for the measured values of fuel consumption and raw exhaust emissions of the four representative cycles are summarized.

As can be seen in Table 5.3, fuel consumption, carbon monoxide (CO), hydrocarbons (HC), and soot emissions are significantly reduced, while nitrogen oxide (NOx) emissions are slightly increased. Note that the optimal rotational speed of the ICE is in the lower speed range, which has higher specific NOx emissions and therefore increase the NOx emissions. Nevertheless, the vehicle's after treatment system is capable of compensating the slightly increased raw NOx emissions.

It can be observed that significant SoC deviations mainly occur for $Q_m = 0.1$, which is caused by the small state penalty that focuses more on maximizing the powertrain efficiency than minimizing the SoC deviations. Nevertheless, the capacity limits of the battery are not even closely reached, since the overall energy conversion is large compared to the deviations, and two of the four test runs even increased the SoC in average.

5.2.3.3 Benefit for a Cyclically Working Wheel Loader Using CD

Filla (2013) analyzed different work cycles of wheel loaders in order to optimize the path trajectories to decrease fuel consumption and to increase productivity, respectively. The cycle detection targets for the same aims, but only based on the past load trajectory, since the applied cycles are unknown in advance. In Fig. 5.4, the influence on the system behavior due to enabled (subscript extension CD) and disabled cycle detection is depicted in more detail for the measurements of cycle number 3.

Subplot one in Fig. 5.4 shows the rotational speed, while marked points (a) emphasize that due to the CD, the speed is increased prior to the load peak. This reduces the need of the electrical system to support any speed changes, which can be seen in Subplot three, where the ISG torques are depicted. The SoC trajectories are shown in the fourth subplot, where the marked points (b) clearly show the difference. After the second recurrence, the cycle is detected and the master MPC optimizes across the predicted load trajectory, which enforces to optimally use the energy storage capabilities, minimizes fuel consumption as well as exhaust emissions and fully considers phlegmatization. Any slightly larger deviation from the demand SoC has therefore less influence on the cost function than the efficiency improvement. It is important

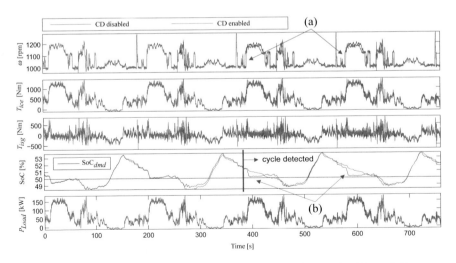

Fig. 5.4 Influence of an enabled cycle detection on the powertrain operation ($Q_m = 0.1$)

to mention that the cycle detection, in principle, needs at least two or more cycles to detect a cycle and reduces noise emissions due to the smoothed operation of the powertrain.

Another way to express the benefit is to calculate the rate of electrical and overall energy conversion, which follows for cycle number 3 to 28.14 % with disabled CD and 25.21 % with enabled CD. In other words, almost 3 % less electrical system usage can be achieved if the information available from the past load trajectory is used accordingly in the control concept. Nevertheless, the reduction in fuel consumption and exhaust emissions is the main aim of the control concept, and keeping the SoC at the demand SoC value is only of secondary importance. On this account, there is a great importance on the accuracy of the SoC value, though.

Chapter 6
Conclusion and Outlook

In this book, the hybridization of non-road vehicles is discussed in terms of the control aspect. The book is motivated by the continuously increasing legislative regulations of non-road mobile machinery to decrease exhaust emissions and fuel consumption. Two topics are mainly relevant to be discussed in this context: the energy management system of the hybrid electric powertrain and the mandatory accurate estimation of the battery state of charge during operation.

First, the generic methodology for nonlinear system identification of battery models in the context of accurate real-time SoC estimation is discussed. The SoC is not measurable online and needs to be estimated during operation, which in case of non-road vehicles is difficult due to the high dynamic usage of the electrical system. In order to obtain a precise battery model for different battery cell chemistries and different temperatures, the data-based local model network approach is used to model the battery cell terminal voltage. The LMN consists of local linear models, which are interpolated to obtain the global nonlinear model output, while the LMN structure is iteratively built by the automatic LOLIMOT algorithm. Battery cells have strong nonlinear effects acting on the voltage, which need to be considered in the model structure. To this end, SoC, current, temperature, relaxation, and hysteresis effects are integrated by corresponding inputs in the LMN structure, which is enhanced by a prepartitioned network to achieve a physically appropriate network. A significant increase in model accuracy results from optimal model-based design of experiments, where a model of the battery is used to optimize the excitation signal of battery cell tests. The high dynamic excitation signal consists of sufficient high currents, which are necessarily required for non-road applications. Furthermore, a real load cycle analysis is made to especially consider frequently used load ranges in operation. The results showed that a battery cell model accuracy with less than 3 % NRMSE could be achieved, while currents above $20C$ were applied to the cell. Result reproducibility and comparability is guaranteed by the proposed measurement procedure.

Based on the battery cell model, different battery module models are built, while the results showed that the consideration of the additional internal resistance due to the cell connections can obviously increase the model quality significantly. Nevertheless, the results also showed that the disregard of the battery module internal

© The Author(s) 2016
J. Unger et al., *Energy Efficient Non-Road Hybrid Electric Vehicles*,
SpringerBriefs in Applied Sciences and Technology,
DOI 10.1007/978-3-319-29796-5_6

resistance also achieves reasonable accuracy, which leads to the fact that if no battery module is available for measurements, the battery module model can nevertheless be built and used for principle analyses. A NRMSE of significantly less than 1.4% for the battery module models could be achieved. The obtained battery models could further be used in a SoC estimator based on the theory of Kalman filter. Due to the LMN models, a fuzzy observer is used to dynamically estimate the SoC during operation, which could achieve an accuracy below 5% depending on the used filter tuning. The results obtained with the fuzzy observer were compared to the SoC estimation provided by the battery management system of the battery module, which showed that the assumption of an inaccurate SoC estimation of the BMS after some time is justified.

Second, an energy management system is presented for a non-road parallel hybrid electric powertrain that considers physical constraints and the future load demand to achieve an optimal control. A cascaded model predictive control concept provides the possibilities to implement the mentioned requirements. The future load demand is in advance unknown, due to which two data-based methodologies are introduced, which predict the disturbance trajectories of the vehicle with sufficient accuracy to be used in the control concept. To this end, a short-term load trajectory prediction based on the Bayesian inference provides a load trajectory for the fast inner control loop, while a cycle detection based on correlation analysis is used to detect recurrent cycles in the past load signal. The strong nonlinear behavior of the electrical system is fully integrated by an iterative optimization of the master control loop.

Simulations and real-time test bed measurements verify the feasibility of the concept on the application example of a wheel loader, while also a theoretical proof of stability is given. The special strategies of downspeeding and phlegmatization have a significant influence on powertrain dynamics, which could be compensated by the proposed controller. At the expense of slightly increased NOx exhaust emissions, CO, HC, and soot emissions as well as fuel consumption are significantly reduced. The results showed furthermore that a change in the MPC penalties can easily implement different hybrid strategies on the powertrain, due to which an easy portability to other vehicles is given.

Future work needs to be focused on an experimental vehicle, which must prove the overall methodology of energy and battery management system. Furthermore, the controller can be improved by integrating material behavior and penetration force of the wheels in the process model (see e.g., Ning and Liu 2013; Blouin et al. 2001).

References

Abu-Sharkh S, Doerffel D (2004) Rapid test and non-linear model characterisation of solid-state lithium-ion batteries. J Power Sources 130(1–2):266–274

Allgöwer F, Zheng A (2000) Nonlinear model predictive control, vol 26. Birkhäuser, Basel

Andrea D (2010) Battery management systems for large lithium ion battery packs. Artech House, Norwood

Babuska R (1998) Fuzzy modeling for control. Kluwer Academic Publishers, Boston

Back M (2005) Prädiktive Antriebsregelung zum energieoptimalen Betrieb von Hybridfahrzeugen. Univ.-Verlag Karlsruhe, Online. http://www.ksp.kit.edu/

Banach S (1922) Sur les opérations dans les ensembles abstraits et leur application aux équations intégrales. Fundam Math 3(1):133–181

Bayindir KC, Gözükücük MA, Teke A (2011) A comprehensive overview of hybrid electric vehicle: powertrain configurations, powertrain control techniques and electronic control units. Energy Convers Manag 52(2):1305–1313. doi:10.1016/j.enconman.2010.09.028

Bentley W (1997) Cell balancing considerations for lithium-ion battery systems. In: Twelfth annual battery conference on applications and advances. IEEE, pp 223–226

Bergsten P, Palm R, Driankov D (2002) Observers for Takagi–Sugeno fuzzy systems. IEEE Trans Syst Man Cybern Part B: Cybern 32(1):114–121

Bernardi D, Go J (2011) Analysis of pulse and relaxation behavior in lithium-ion batteries. J Power Sources 196(1):412–427

Bhangu BS, Bentley P, Stone DA, Bingham CM (2005) Nonlinear observers for predicting state-of-charge and state-of-health of lead-acid batteries for hybrid-electric vehicles. IEEE Trans Veh Technol 54(3):783–794

Billings S, Aguirre LA (1995) Effects of the sampling time on the dynamics and identification of nonlinear models. Int J Bifurc Chaos 5(06):1541–1556

Blom HA, Bar-Shalom Y (1988) The interacting multiple model algorithm for systems with Markovian switching coefficients. IEEE Trans Autom Control 33(8):780–783

Blouin S, Hemami A, Lipsett M (2001) Review of resistive force models for earthmoving processes. J Aerosp Eng 14(3):102–111

Borhan HA, Zhang C, Vahidi A, Phillips AM, Kuang ML, Di Cairano S (2010) Nonlinear model predictive control for power-split hybrid electric vehicles. In: 49th IEEE conference on decision and control (CDC). IEEE, pp 4890–4895

Box GE, Tiao GC (2011) Bayesian inference in statistical analysis, vol 40. Wiley, New York

Camacho E, Bordons C (1999) Model predictive control. Springer, Berlin

Cao J, Schofield N, Emadi A (2008) Battery balancing methods: a comprehensive review. In: Vehicle power and propulsion conference (VPPC'08), IEEE, pp 1–6

© The Author(s) 2016
J. Unger et al., *Energy Efficient Non-Road Hybrid Electric Vehicles*,
SpringerBriefs in Applied Sciences and Technology,
DOI 10.1007/978-3-319-29796-5

Charkhgard M, Farrokhi M (2010) State-of-charge estimation for lithium-ion batteries using neural networks and EKF. IEEE Trans Ind Electron 57(12):4178–4187

Chaturvedi N, Klein R, Christensen J, Ahmed J, Kojic A (2010) Algorithms for advanced battery management systems: modeling, estimation, and control challenges for lithium-ion batteries. IEEE Control Syst Mag 30(3):49–68

Chen G, Xie Q, Shieh LS (1998) Fuzzy kalman filtering. Inf Sci 109(1–4):197–209. http://dx.doi.org/10.1016/S0020-0255(98)10002--6

Chen M, Rincon-Mora G (2006) Accurate electrical battery model capable of predicting runtime and IV performance. IEEE Trans Energy Convers 21(2):504–511

Chen Z, Qiu S, Masrur MA, Murphey YL (2011) Battery state of charge estimation based on a combined model of extended kalman filter and neural networks. In: The 2011 international joint conference on neural networks (IJCNN). IEEE, pp 2156–2163

Cohn D (1996) Neural network exploration using optimal experiment design. Neural Netw 9(6):1071–1083

D'Agostino RB (1986) Goodness-of-fit-techniques, vol 68. CRC Press, Boca Raton

Daowd M, Omar N, Van Den Bossche P, Van Mierlo J (2011) Passive and active battery balancing comparison based on matlab simulation. In: Vehicle power and propulsion conference (VPPC'11). IEEE, pp 1–7

De Sá JM (2007) Applied statistics: using SPSS, statistica, MATLAB, and R, 2nd edn. Springer, Heidelberg

Deflorian M, Klöpper F (2009) Design of dynamic experiments. Design of experiments (DoE) in engine development, vol 4, pp 31–40

Desai C, Williamson SS (2009) Comparative study of hybrid electric vehicle control strategies for improved drivetrain efficiency analysis. In: Electrical power & energy conference (EPEC). IEEE, pp 1–6

Do DV, Forgez C, El Kadri Benkara K, Friedrich G (2009) Impedance observer for a li-ion battery using kalman filter. IEEE Trans Veh Technol 58(8):3930–3937

Doyle M, Fuller T, Newman J (1993) Modeling of galvanostatic charge and discharge of the lithium/polymer/insertion cell. J Electrochem Soc 140(6):1526–1533

Dubarry M, Vuillaume N, Liaw BY (2008) From li-ion single cell model to battery pack simulation. In: IEEE international conference on control applications (CCA). IEEE, pp 708–713

Dubarry M, Vuillaume N, Liaw BY (2009) From single cell model to battery pack simulation for li-ion batteries. J Power Sources 186(2):500–507

Fekri S, Assadian F (2012) The role and use of robust multivariable control in hybrid electric vehicle energy management-part i: an overview. In: IEEE international conference on control applications (CCA). IEEE, pp 303–309

Filla R (2009) Hybrid power systems for construction machinery: aspects of system design and operability of wheel loaders. In: International mechanical engineering congress and exposition. ASME, pp 611–620

Filla R (2013) Optimizing the trajectory of a wheel loader working in short loading cycles. In: The 13th scandinavian international conference on fluid power (SICFP2013), pp 307–317

Forgez C, Do DV, Friedrich G, Morcrette M, Delacourt C (2010) Thermal modeling of a cylindrical $LiFePO_4$/graphite lithium-ion battery. J Power Sources 195(9):2961–2968

Frank B, Skogh L, Alaküla M (2012) On wheel loader fuel efficiency difference due to operator behaviour distribution. In: 2nd international commercial vehicle technology symposium (CVT)

Ganji B, Kouzani AZ (2011) A look-ahead road grade determination method for hevs. Electrical engineering and control. Springer, New York, pp 703–711

Gao L, Liu S, Dougal R (2002) Dynamic lithium-ion battery model for system simulation. IEEE Trans Compon Packag Technol 25(3):495–505

Gao W, Porandla SK (2005) Design optimization of a parallel hybrid electric powertrain. In: Vehicle power and propulsion conference (VPPC'05). IEEE, pp 6

Gao Y (2012) Existence and uniqueness theorem on uncertain differential equations with local Lipschitz condition. J Uncertain Syst 6(3):223–232

Gilbert EG, Tan KT (1991) Linear systems with state and control constraints: The theory and application of maximal output admissible sets. IEEE Trans Autom Control 36(9):1008–1020

Gomez J, Nelson R, Kalu EE, Weatherspoon MH, Zheng JP (2011) Equivalent circuit model parameters of a high-power li-ion battery: Thermal and state of charge effects. J Power Sources 196(10):4826–4831

González A, Adam E, Marchetti J (2008) Conditions for offset elimination in state space receding horizon controllers: A tutorial analysis. Chem Eng Process: Process Intensif 47(12):2184–2194

Goodwin G, Payne R (1977) Dynamic system identification: experiment design and data analysis. academic press inc., New York

Gregorcic G, Lightbody G (2007) Local model network identification with gaussian processes. IEEE Trans Neural Netw 18(5):1404–1423

Gunderson DS (2010) Handbook of mathematical induction: theory and applications. AMC 10:12

Hametner C, Jakubek S (2007) Neuro-fuzzy modelling using a logistic discriminant tree. In: American control conference. IEEE, pp 864–869

Hametner C, Jakubek S (2013) State of charge estimation for lithium ion cells: design of experiments, nonlinear identification and fuzzy observer design. J Power Sources 238:413–421. doi:10.1016/j.jpowsour.2013.04.040

Hametner C, Unger J, Jakubek S (2012) Local model network based dynamic battery cell model identification. In: 12th WSEAS international conference on robotics. Control and manufacturing technology, WSEAS, pp 116–123

Hametner C, Stadlbauer M, Deregnaucourt M, Jakubek S (2013a) Incremental optimal process excitation for online system identification based on evolving local model networks. Math Comput Modell Dyn Syst 19(6):505–525

Hametner C, Stadlbauer M, Deregnaucourt M, Jakubek S, Winsel T (2013b) Optimal experiment design based on local model networks and multilayer perceptron networks. Eng Appl Artif Intell 26(1):251–261

Hametner C, Prochazka W, Suljanovic A, Jakubek S (2014) Model based lithium ion cell ageing data analysis. In: IEEE international conference on fuzzy systems (FUZZ-IEEE). IEEE, pp 962–967

Han J, Kim D, Sunwoo M (2009) State-of-charge estimation of lead-acid batteries using an adaptive extended kalman filter. J Power Sources 188(2):606–612

Hartmann B, Moll J, Nelles O, Fritzen C (2011) Hierarchical local model trees for design of experiments in the framework of ultrasonic structural health monitoring. In: IEEE international conference on control applications. IEEE, pp 1163–1170

He H, Xiong R, Zhang X, Sun F, Fan J (2011) State-of-charge estimation of the lithium-ion battery using an adaptive extended kalman filter based on an improved thevenin model. IEEE Trans Veh Technol 60(4):1461–1469

Helm S, Kozek M, Jakubek S (2012) Combustion torque estimation and misfire detection for calibration of combustion engines by parametric Kalman filtering. IEEE Trans Ind Electron 59(11):4326–4337

Hentunen A, Lehmuspelto T, Suomela J (2011) Electrical battery model for dynamic simulations of hybrid electric vehicles. In: Vehicle power and propulsion conference (VPPC'11). IEEE, pp 1–6

Hoffmann F, Nelles O (2001) Genetic programming for model selection of tsk-fuzzy systems. Inf Sci 136(1):7–28

Hu X, Lin S, Stanton S (2010a) A novel thermal model for HEV/EV battery modeling based on CFD calculation. In: Energy conversion congress and exposition (ECCE). IEEE, pp 893–900

Hu X, Sun F, Zou Y (2010b) Estimation of state of charge of a lithium-ion battery pack for electric vehicles using an adaptive Luenberger observer. Energies 3(9):1586–1603

Hu Y, Yurkovich S (2012) Battery cell state-of-charge estimation using linear parameter varying system techniques. J Power Sources 198:338–350

Hu Y, Yurkovich B, Yurkovich S, Guezennec Y (2009a) Electro-thermal battery modeling and identification for automotive applications. In: Dynamic systems and control conference. ASME, pp 233–240

Hu Y, Yurkovich S, Guezennec Y, Yurkovich B (2009b) A technique for dynamic battery model identification in automotive applications using linear parameter varying structures. Control Eng Prac 17(10):1190–1201

Hu Y, Yurkovich S, Guezennec Y, Yurkovich B (2011) Electro-thermal battery model identification for automotive applications. J Power Sources 196(1):449–457

Hulnhagen T, Dengler I, Tamke A, Dang T, Breuel G (2010) Maneuver recognition using probabilistic finite-state machines and fuzzy logic. In: Intelligent vehicles symposium (IV). IEEE, pp 65–70

Istratescu VI (2001) Fixed point theory: an introduction, vol 7. Springer, New York

Jackey R, Saginaw M, Sanghvi P, Gazzarri J, Huria T, Ceraolo M (2013) Battery model parameter estimation using a layered technique: An example using a lithium iron phosphate cell. SAE Technical Paper 2013-01-1547 doi:10.4271/2013-01-1547

Jakubek S, Hametner C (2009) Identification of neurofuzzy models using GTLS parameter estimation. IEEE Trans Syst Man Cybern Part B: Cybern 39(5):1121–1133

Jakubek S, Keuth N (2006) A local neuro-fuzzy network for high-dimensional models and optimization. Eng Appl Artif Intell 19(6):705–717

Jeon DH, Baek SM (2011) Thermal modeling of cylindrical lithium ion battery during discharge cycle. Energy Convers Manag 52(8):2973–2981

Jing Z, Luo A, Tomizuka M (1998) A stochastic fuzzy neural network for nonlinear dynamic systems. Int J Intell Control Syst 3(2):193–203

Johnson VH, Wipke KB, Rausen DJ (2000) Hev control strategy for real-time optimization of fuel economy and emissions. SAE Technical Paper 2000-01-1543 doi:10.4271/2000-01-1543

Jwo DJ, Tseng CH (2009) Fuzzy adaptive interacting multiple model unscented kalman filter for integrated navigation. In: control applications (CCA) & intelligent control (ISIC). IEEE, pp 1684–1689

Katrasnik T (2009) Analytical framework for analyzing the energy conversion efficiency of different hybrid electric vehicle topologies. Energy Convers Manag 50(8):1924–1938

Katsargyri GE, Kolmanovsky IV, Michelini J, Kuang ML, Phillips AM, Rinehart M, Dahleh MA (2009) Optimally controlling hybrid electric vehicles using path forecasting. In: American control conference, institute of electrical and electronics engineers, pp 4613–4617

Keerthi S, Gilbert E (1987) Computation of minimum-time feedback control laws for discrete-time systems with state-control constraints. IEEE Trans Autom Control 32(5):432–435

Kim IS (2006) The novel state of charge estimation method for lithium battery using sliding mode observer. J Power Sources 163(1):584–590

Kim J, Shin J, Jeon C, Cho B (2011) High accuracy state-of-charge estimation of li-ion battery pack based on screening process. In: 26th annual IEEE applied power electronics conference and exposition (APEC). IEEE, pp 1984–1991

Kim J, Shin J, Chun C, Cho B (2012) Stable configuration of a Li-ion series battery pack based on a screening process for improved voltage/SOC balancing. IEEE Trans Power Electron 27(1):411–424

Klein R, Chaturvedi N, Christensen J, Ahmed J, Findeisen R, Kojic A (2010) State estimation of a reduced electrochemical model of a lithium-ion battery. In: American control conference. IEEE, pp 6618–6623

König O, Gregorčič G, Jakubek S (2013) Model predictive control of a DC–DC converter for battery emulation. Control Eng Prac 21(4):428–440

Konig O, Prochart G, Hametner C, Jakubek S (2014) Battery emulation for power-hil using local model networks and robust impedance control. IEEE Trans Ind Electron 61(2):943–955

Kroeze R, Krein P (2008) Electrical battery model for use in dynamic electric vehicle simulations. In: Power electronics specialists conference. IEEE, pp 1336–1342

Kwon TS, Lee SW, Sul SK, Park CG, Kim NI, Kang Bi, Hong Ms (2010) Power control algorithm for hybrid excavator with supercapacitor. IEEE Trans Ind Appl 46(4):1447–1455

Lee S, Kim J, Lee J, Cho B (2008) State-of-charge and capacity estimation of lithium-ion battery using a new open-circuit voltage versus state-of-charge. J Power Sources 185(2):1367–1373

Lee SS, Kim TH, Hu SJ, Cai WW, Abell JA (2010) Joining technologies for automotive lithium-ion battery manufacturing: A review. In: International manufacturing science and engineering conference. ASME, pp 541–549

Lee YS, Cheng MW (2005) Intelligent control battery equalization for series connected lithium-ion battery strings. IEEE Trans Ind Electron 52(5):1297–1307

Lendek Z, Babuska R, De Schutter B (2009) Stability of cascaded fuzzy systems and observers. IEEE Trans Fuzzy Syst 17(3):641–653

Lin CC, Peng H, Grizzle JW, Kang JM (2003) Power management strategy for a parallel hybrid electric truck. IEEE Trans Control Syst Technol 11(6):839–849

Lin CC, Jeon S, Peng H, Moo Lee J (2004) Driving pattern recognition for control of hybrid electric trucks. Veh Syst Dyn 42(1–2):41–58

Lindenkamp IN, Tilch DIB (2012) Reducing of exhaust emissions from diesel hybrid vehicles. MTZ Worldw 73(7–8):64–69

Ljung L (1998) System identification. Springer, New York

Lorenz A, Kozek M (2007) Automatic cycle border detection for a statistic evaluation of the loading process of earth-moving vehicles. In: Commercial vehicle engineering congress and exhibition, pp 08–12. doi:10.4271/2007-01-4191

Luenberger D, Ye Y (2008) Linear and nonlinear programming, 3rd edn. Springer, New York

Lyshevski SE (2000) Energy conversion and optimal energy management in diesel-electric drivetrains of hybrid-electric vehicles. Energy Conversn Manag 41(1):13–24

Magnus J, Neudecker H (1988) Matrix differential calculus with applications in statistics and econometrics, 2nd edn. Wiley, Chichester

Marques De Sá JP (2003) Applied statistics using SPSS, STATISTICA and MATLAB. Springer, New York

Mayne DQ, Rawlings JB, Rao CV, Scokaert PO (2000) Constrained model predictive control: stability and optimality. Automatica 36(6):789–814

Mayr CH, Fleck A, Jakubek S (2011a) Hybrid powertrain control using optimization and cycle based predictive control algorithms. In: 9th IEEE international conference on control and automation (ICCA). IEEE, pp 937–944

Mayr CH, Hametner C, Kozek M, Jakubek S (2011b) Piecewise quadratic stability analysis for local model networks. In: IEEE International conference on control applications (CCA). IEEE, pp 1418–1424

Mazor E, Averbuch A, Bar-Shalom Y, Dayan J (1998) Interacting multiple model methods in target tracking: a survey. IEEE Trans Aerosp Electron Syst 34(1):103–123

Montazeri-Gh M, Ahmadi A, Asadi M (2008) Driving condition recognition for genetic-fuzzy hev control. In: 3rd international workshop on genetic and evolving systems (GEFS). IEEE, pp 65–70

Moura SJ, Fathy HK, Callaway DS, Stein JL (2011) A stochastic optimal control approach for power management in plug-in hybrid electric vehicles. IEEE Trans Control Syst Technol 19(3):545–555

Murray-Smith R, Johansen T (1997) Multiple model approaches to nonlinear modelling and control. CRC Press, Boca Raton

Nelles O (2001) Nonlinear system identification: from classical approaches to neural networks and fuzzy models. Springer, Berlin

Nelles O, Isermann R (1996) Basis function networks for interpolation of local linear models. In: Proceedings of the 35th IEEE decision and control, vol 1. IEEE, pp 470–475

Ning Y, Liu XH (2013) Research on the resistance acting on the bucket during shovelling. Adv Mater Res 787:778–781

Nüesch T, Wang M, Isenegger P, Onder CH, Steiner R, Macri-Lassus P, Guzzella L (2014) Optimal energy management for a diesel hybrid electric vehicle considering transient PM and quasi-static NOx emissions. Control Eng Prac 29:266–276. doi:10.1016/j.conengprac.2014.01.020

Ogata K (1995) Discrete-time control systems, vol 8, 2nd edn. Prentice-Hall, Englewood Cliffs

Onda K, Ohshima T, Nakayama M, Fukuda K, Araki T (2006) Thermal behavior of small lithium-ion battery during rapid charge and discharge cycles. J Power sources 158(1):535–542

Pattipati B, Sankavaram C, Pattipati KR (2010) System identification and estimation framework for pivotal automotive battery management system characteristics. IEEE Trans Syst Man Cybern, Part C: Appl Rev pp 1–16. doi:10.1109/TSMCC.2010.2089979

Payri F, Guardiola C, Pla B, Blanco-Rodriguez D (2014) A stochastic method for the energy management in hybrid electric vehicles. Control Eng Prac 29:257–265. doi:10.1016/j.conengprac.2014.01.004

Pisu P, Rizzoni G (2007) A comparative study of supervisory control strategies for hybrid electric vehicles. IEEE Trans Control Syst Technol 15(3):506–518

Plett G (2004a) Extended Kalman filtering for battery management systems of LiPB-based HEV battery packs: Part 1. Background. J Power Sources 134(2):252–261

Plett G (2004b) Extended Kalman filtering for battery management systems of LiPB-based HEV battery packs: Part 2 Modeling and identification. J Power Sources 134(2):262–276

Plett G (2004c) Extended Kalman filtering for battery management systems of LiPB-based HEV battery packs: Part 3 State and parameter estimation. J Power Sources 134(2):277–292

Plett GL (2004d) High-performance battery-pack power estimation using a dynamic cell model. IEEE Trans Veh Technol 53(5):1586–1593

Poursamad A, Montazeri M (2008) Design of genetic-fuzzy control strategy for parallel hybrid electric vehicles. Control Eng Prac 16(7):861–873

Powell B (1979) A dynamic model for automotive engine control analysis. In: 18th IEEE conference on decision and control including the symposium on adaptive processes, vol 18. IEEE, pp 120–126

Pronzato L (2008) Optimal experimental design and some related control problems. Automatica 44(2):303–325

Quoc TD, Savorgnan C, Diehl M (2012) Real-time sequential convex programming for optimal control applications. Modeling, simulation and optimization of complex processes. Springer, New York

Rawlings JB, Mayne DQ (2009) Model predictive control: theory and design. Nob Hill Pub, Madison

Richter S, Jones CN, Morari M (2009) Real-time input-constrained MPC using fast gradient methods. In: 48th IEEE conference on decision and control held jointly with the 28th chinese control conference (CDC/CCC). IEEE, pp 7387–7393

Richter S, Mariéthoz S, Morari M (2010) High-speed online MPC based on a fast gradient method applied to power converter control. In: American control conference (ACC). IEEE, pp 4737–4743

Rojas OJ, Goodwin GC, Seron MM, Feuer A (2004) An SVD based strategy for receding horizon control of input constrained linear systems. Int J Robust Nonlinear Control 14:1207–1226. doi:10.1002/rnc.940

Santhanagopalan S, White R (2006) Online estimation of the state of charge of a lithium ion cell. J Power Sources 161(2):1346–1355

Sciarretta A, Guzzella L (2007) Control of hybrid electric vehicles. IEEE Control Syst 27(2):60–70

Sciarretta A, Back M, Guzzella L (2004) Optimal control of parallel hybrid electric vehicles. IEEE Trans Control Syst Technol 12(3):352–363

Sen C, Kar NC (2009) Battery pack modeling for the analysis of battery management system of a hybrid electric vehicle. In: Vehicle power and propulsion conference (VPPC'09). IEEE, pp 207–212

Senthil R, Janarthanan K, Prakash J (2006) Nonlinear state estimation using fuzzy kalman filter. Ind Eng Chem Res 45(25):8678–8688

Seyr M, Jakubek S (2007) Dynamic trajectory generation via numerical multi-objective optimisation. In: American control conference. IEEE, pp 3336–3341

Simon D (2003) Kalman filtering for fuzzy discrete time dynamic systems. Appl Soft Comput 3(3):191–207

Simon D, Lofberg J, Glad T (2012) Reference tracking MPC using terminal set scaling. In: IEEE 51st annual conference on decision and control (CDC). IEEE, pp 4543–4548

Smith K, Wang C (2006) Power and thermal characterization of a lithium-ion battery pack for hybrid-electric vehicles. J Power Sources 160(1):662–673

Smith K, Rahn C, Wang C (2010) Model-based electrochemical estimation and constraint management for pulse operation of lithium ion batteries. IEEE Trans Control Syst Technol 18(3):654–663

Song H, Shin V, Jeon M (2012) Mobile node localization using fusion prediction-based interacting multiple model in cricket sensor network. IEEE Trans Ind Electron 59(11):4349–4359

Stadlbauer M, Hametner C, Jakubek S (2011a) Analytic model based design of experiments for non-linear dynamic systems with constraints. In: Proceedings of the IASTED international conference in control and applications, vol 20

Stadlbauer M, Hametner C, Jakubek S, Winsel T (2011b) Analytic multilayer perceptron based experiment design for nonlinear systems. In: Proceedings of the eighth IFAC world congress, pp 4332–4337

Sun F, Hu X, Zou Y, Li S (2011) Adaptive unscented Kalman filtering for state of charge estimation of a lithium-ion battery for electric vehicles. Energy 36(5):3531–3540

Sun F, Xiong R, He H, Li W, Aussems J (2012a) Model-based dynamic multi-parameter method for peak power estimation of lithium-ion batteries. Appl Energy 96:378–386

Sun H, Wang X, Tossan B, Dixon R (2012b) Three-dimensional thermal modeling of a lithium-ion battery pack. J Power Sources 206:349–356

Tanaka K, Ikeda T, Wang HO (1998) Fuzzy regulators and fuzzy observers: relaxed stability conditions and LMI-based designs. IEEE Trans Fuzzy Syst 6(2):250–265

Tang X, Zhang X, Koch B, Frisch D (2008) Modeling and estimation of Nickel Metal Hydride battery hysteresis for SOC estimation. In: IEEE international conference on prognostics and health management. IEEE, pp 1–12

The European Parliament and Council of the European Union (1997) Directive 97/68/EC of the European Parliament and of the Council of 16 December 1997 on the approximation of the laws of the Member States relating to measures against the emission of gaseous and particulate pollutants from internal combustion engines to be installed in non-road mobile machinery

The European Parliament and Council of the European Union (2014) Proposal for a regulation of the European Parliament and of the Council on requirements relating to emission limits and type-approval for internal combustion engines for non-road mobile machinery, 2014/0268/COD

Tondel P, Johansen TA (2002) Complexity reduction in explicit linear model predictive control. In: 15th Triennial world congress of the international federation of automatic control

Unger J (2015) Energy and battery management for non-road hybrid electric vehicles. Dissertation, Vienna University of Technology

Unger J, Hametner C, Jakubek S (2012a) Optimal model based design of experiments applied to high current rate battery cells. In: IEEE conference on electrical systems for aircraft, railway and ship propulsion. IEEE, pp 1–6

Unger J, Kozek M, Jakubek S (2012b) Reduced order optimization for model predictive control using principal control moves. J Process Control 22(1):272–279

Unger J, Hametner C, Jakubek S, Quasthoff M (2014) A novel methodology for non-linear system identification of battery cells used in non-road hybrid electric vehicles. J Power Sources 269:883–897

Unger J, Kozek M, Jakubek S (2015) Nonlinear model predictive energy management controller with load and cycle prediction for non-road HEV. Control Eng Prac 36:120–132. doi:10.1016/j.conengprac.2014.12.001

Vasebi A, Partovibakhsh M, Bathaee SMT (2007) A novel combined battery model for state-of-charge estimation in lead-acid batteries based on extended kalman filter for hybrid electric vehicle applications. J Power Sources 174(1):30–40

Vasebi A, Bathaee S, Partovibakhsh M (2008) Predicting state of charge of lead-acid batteries for hybrid electric vehicles by extended Kalman filter. Energy Convers Manag 49(1):75–82

Verbrugge M, Tate E (2004) Adaptive state of charge algorithm for nickel metal hydride batteries including hysteresis phenomena. J Power Sources 126(1–2):236–249

Vidyasagar M (2002) Nonlinear systems analysis, vol 42. SIAM, Philadelphia

Wang J, Chen Q (2005) Stochastic fuzzy neural network and its robust parameter learning algorithm. Adv Neural Netw, pp 199–221

Wang J, Xu L, Guo J, Ding L (2009) Modelling of a battery pack for electric vehicles using a stochastic fuzzy neural network. Proc Inst Mech En Part D: J Automob Eng 223(1):27–35

Wang L (1994) Adaptive fuzzy systems and control: Design and stability analysis. Prentice Hall, Englewood Cliffs

Wang L (2009) Model predictive control system design and implementation using Matlab, 1st edn. Springer, New York

Wang Y, Boyd S (2010) Fast model predictive control using online optimization. IEEE Trans Control Syst Technol 18(2):267–278

Watrin N, Bouquain D, Blunier B, Miraoui A (2011) Multiphysical lithium-based battery pack modeling for simulation purposes. In: Vehicle power and propulsion conference (VPPC'11). IEEE, pp 1–5

Weibel M, Schmeißer V, Hofmann F (2014) Model-based approaches to exhaust aftertreatment system development. Urea-SCR technology for deNOx after treatment of diesel exhausts. Springer, New York, pp 691–707

Weicker P (2013) A systems approach to Lithium-Ion battery management. Artech house, Boston

Xiao Q, Wang Q, Zhang Y (2008) Control strategies of power system in hybrid hydraulic excavator. Autom Constr 17(4):361–367

Xu L, Wang J, Chen Q (2012) Kalman filtering state of charge estimation for battery management system based on a stochastic fuzzy neural network battery model. Energy Convers Manag 53(1):33–39

Yan F, Wang J, Huang K (2012) Hybrid electric vehicle model predictive control torque-split strategy incorporating engine transient characteristics. IEEE Trans Veh Technol 61(6):2458–2467

Yuan S, Wu H, Yin C (2013) State of charge estimation using the extended Kalman filter for battery management systems based on the ARX battery model. Energies 6(1):444–470. doi:10.3390/en6010444

Zeilinger MN, Jones CN, Morari M (2011) Real-time suboptimal model predictive control using a combination of explicit MPC and online optimization. IEEE Trans Autom Control 56(7):1524–1534